Computational Intelligence for Human Action Recognition

Chapman & Hall/CRC Computational Intelligence and Its Applications

Series Editor: Siddhartha Bhattacharyya

Intelligent Copyright Protection for Images
Subhrajit Sinha Roy, Abhishek Basu, Avik Chattopadhyay

Emerging Trends in Disruptive Technology Management for Sustainable Development
Rik Das, Mahua Banerjee, Sourav De

Computational Intelligence for Human Action Recognition
Sourav De, Paramartha Dutta

For more information about this series please visit:
https://www.crcpress.com/Chapman--HallCRC-Computational-Intelligence-and-Its-Applications/book-series/CIAFOCUS

Computational Intelligence for Human Action Recognition

Edited by
Sourav De
Paramartha Dutta

CRC Press
Taylor & Francis Group
Boca Raton London New York

CRC Press is an imprint of the
Taylor & Francis Group, an **informa** business
A CHAPMAN & HALL BOOK

First edition published 2021
by CRC Press
6000 Broken Sound Parkway NW, Suite 300, Boca Raton, FL 33487-2742

and by CRC Press
2 Park Square, Milton Park, Abingdon, Oxon, OX14 4RN

© 2021 Taylor & Francis Group, LLC

CRC Press is an imprint of Taylor & Francis Group, LLC

ISBN: 978-0-367-18443-8 (hbk)
ISBN: 978-0-367-53943-6 (pbk)
ISBN: 978-0-429-06148-6 (ebk)

Typeset in LMRoman
by Nova Techset Private Limited, Bengaluru & Chennai, India

Visit the Taylor & Francis Web site at
http://www.taylorandfrancis.com

and the CRC Press Web site at
http://www.crcpress.com

*Dr. Sourav De dedicates this book to his respected
mentor Prof. (Dr.) Siddhartha Bhattacharyya*

*Prof. (Dr.) Paramartha Dutta dedicates this book to his parents
Late Arun Kanti Dutta and Late Bandana Dutta.*

Contents

ASIT BARMAN, SANKHAYAN CHODHURY, and PARAMARTHA DUTTA

CHAPTER   5 ▪ A Behavioural Model for Persons with
                Autism Based on Relevant Case Study 103

RUDRANATH BANERJEE, SOURAV DE, and SHOUVIK DEY

Preface

Human action recognition is an important task in day-to-day life. When the recognition is done in an automated system, the conventional methods are not effective all the time. Basically for human action recognition, human appearances and motion patterns are classified into different action classes.

Human action recognition (HAR) is a widely studied computer vision problem and has many valuable applications in computer vision such as human-computer interaction, video surveillance, electronic entertainment, health care, patient monitoring, nursing homes, smart homes, etc. The advancement of human action recognition emerged constantly due to the advancement of the imaging techniques and the upgradation of the acquisition devices. Different human gestures can be distinguished from the movement of the human skeleton. The meaning of three words, viz. gesture, action and activity are very close to one other. Basically, gestures signify elementary movements of human body parts to describe meaningful motions of human body. Composition of multiple gestures of a single-person activity makes an action. However, it is a challenging task to recognize actions accurately due to cluttered backgrounds, occlusions, and viewpoint variations, etc. Some of the classical and the state-of the-art approaches for human action recognition are Histogram of Optical Flow (HOF), Histogram of Oriented Gradient (HOG), Space-Time Interest Points (STIP), Motion Interchange Patters (MIP), dense trajectories, etc. The extension of these features to other systems is not very straightforward as these approaches are difficult and time consuming. Different tasks are applied to determine the large part of hand-design features and they may enforce different features. In real life, feature selection is very much problem sensitive as features are task specific. Uncertainty, imprecision, vagueness, are some of criteria associated to these real life human action recognition problems.

The conventional paradigms may not offer accurate solutions to figure out such types of problems. The latest intelligent computing paradigms based on soft computing tools are expected to yield better solutions to handle these problems. Computational intelligence, in all likelihood, is expected to provide the solution platform to handle these issues to a considerable extent. The field of computational intelligence furnishes the computing paradigm which faithfully resolves the uncertainties prevalent in the rapidly evolving computing arena. An intelligent machine inherits the boon of intelligence, artificial albeit, by virtue of the various methodologies encompassing computationally intelligent paradigm encompassing fuzzy and rough set theory, artificial neuro-computing, evolutionary computing, as well as approximate reasoning. Using computational intelligence techniques, the process of feature construction can be done by a set of machines that can learn a hierarchy of features by building high-level features from low-level ones. After training by supervised or unsupervised approaches, the automated system becomes equipped to distinguish competitive performance in visual object recognition, natural language processing, and audio classification tasks.

This include major emerging trends in technology which are supporting the current advancement of the human action recognition with the help of computational intelligence. It also highlights advancement of the conventional approaches in the field of human action recognition. The scope of this book involves proposing novel techniques and reviewing state-of-the-art in the area of machine learning, computer vision, soft computing techniques, etc., and also to relate the same to their applications in human action recognition. The motivation of this endeavour is not only to put forward new ideas in technology innovation but also to analyse the effect of the same in the current context of human action recognition.

The book contains five well versed chapters written by the leading practitioners in the field.

A brief overview of automatic sign language recognition (SLR) is presented in Chapter 1. Different types of modern technologies are introduced in this field as it is a relevant and active research field. Basically, deaf people use sign language for communication. Instead of using the spoken language for communication, they apply visual gestures and signs. These gestures are a combination of movements of face, hand and different parts of the body. This type of language has a broad social impact and very helpful for the hearing-impaired persons. The recent technologies to handle SLR involve Deep Neural Networks, 3D CNN, embedding a CNN into HMM, adaptive pooling which are discussed in this chapter and their challenges and limitations are also illustrated. The recent advancements and future extensions of this area of research are also presented in this chapter.

In the field of the nonverbal communication in vogue in this human society, facial expression plays a pivotal role. During the interpersonal communication by voice, all types of information may fall short while being transferred and this drawback could be overcome by facial expressions. For that reason, facial expression recognition is a growing research requirement in computer vision. The human-computer interaction technology is significantly dependent on human facial expression recognition system and many fields such as medical, computer-aided training and distance education are also dependent on the facial expression recognition system. In Chapter 2, a combined distance-shape-texture signature trio feature is presented for recognizing the expressions from facial images. Distance, shape and texture signatures play a vital role in recognizing emotions. An appearance model is employed to identify the proper landmarks on the faces and the grid of the human face is formed with the help of selected landmark points. The distance signature, shape signature and texture signature are determined from the grid and salient landmark points. After that, they are applied to get a distance-shape-texture signature trio feature to fulfill facial expression recognition task effectively.

One of the important aspects of human-computer interaction in the field of emotion analysis is the affective analysis. Many real-life applications of automatic analysis of human emotions can be found in medical, computational, security and surveillance, monitoring systems, companion robotics, mood-based suggestion system, etc. The expressions of the face maintain a significant amount of characteristic information for automated recognition of human emotions as different forms of human emotions that emanate from the face and have the highest power to differentiate one emotion from another. In Chapter 3, a statistical model of object shape recognition, viz., Active Appearance Model (AAM) is applied to generate landmark points responsible for describing the geometrical position of the facial components. Those points are employed to formulate triangulation description of the shape of the face. Authors of this chapter propose a novel approach using a different ratio of side lengths of a triangle constituting geometrical shape derived through forming triangulation generated out of landmark points indicating axial positions of facial components fitted by an AAM. The multilayer perceptron (MLP), empowered by the side length features is used effectively to segregate facial expression in six basic classes of expressions.

In Chapter 4, an efficient approach for recognition of human emotion from facial video frames is presented. A triangulation mechanism is applied to distinguish human emotion into different basic facial expressions like anger, disgust, fear, happiness, sadness, and surprise. Important regions of the face are considered to get relevant geometric features capable of identifying emotional change of a human being reflected in an image sequence.

Autism Spectrum Disorder (ASD) is a permanent neuro-developmental disorder that occurs in infancy leading to challenges in social communication and behaviour. It captures a wide spectrum ranging from mild to severe. According to recent studies, ASD is a common disorder with approximately 1 out of every 59 individuals suffering from it. The care for such individuals entails lifelong support in terms of medical, social and emotional support. The social uneasiness and emotional vulnerability leading to behavioural unpredictability makes them live life in isolation which poses real challenges for people to understand them and approach them. In Chapter 5, a case study on a behavioural model for autistic persons is proposed with different parameters influencing behaviours such as emotions, likes and dislikes of an individual, and the surrounding environment with the intention of recognising the exact emotions of the children. All these in turn may ensure a positive environment in this regard. A model has been designed with the help of both psychological and technical support.

The editors, on their part, may find their present endeavour meaningful if the readership, specially comprising the research community finds the contents useful for their future research. The editors further desire to avail this opportunity to express their sincere thanks to CRC Press, Taylor and Francis for providing them the opportunity to publish the present treatise.

Sourav De
Cooch Behar, India

Paramartha Dutta
Kolkata, India

Editors

Dr. Sourav De completed his Bachelors in Information Technology from The University of Burdwan, Burdwan, India in 2002. He earned his Masters in Information Technology from West Bengal University of Technology, Kolkata, India in 2005. He completed his PhD in Computer Science and Technology from Indian Institute of Engineering and Technology, Shibpur, Howrah, India in 2015. He is currently an Associate Professor of Computer Science and Engineering Department in Cooch Behar Government Engineering College, West Bengal. Previous to this, he was an Assistant Professor for more than ten years in the Department of Computer Science and Engineering and Information Technology of University Institute of Technology, The University of Burdwan, Burdwan, India. He served as a Junior Programmer in Apices Consultancy Private Limited, Kolkata, India in 2005. He is a co-author of one book and the co-editor of nine books and has more than 41 research publications in internationally reputed journals, international edited books, and international IEEE conference proceedings, and one patent to his credit. He served as a reviewer in several International IEEE conferences and also in several international editorial books. He also served as a reviewer in some reputed international journals, like, Applied Soft Computing, Elsevier, B. V., Knowledge-Based Systems, Computer Methods in Biomechanics and Biomedical Engineering Imaging and Visualization, Inderscience Journals, etc. He has been the member of the organizing and technical program committees of several national and international conferences. He has been invited to different seminars as an expert speaker. His research interests include soft computing, pattern recognition, image processing, and data mining. Dr. De is a member of IEEE, ACM, Computer Science Teachers Association (CSTA), Institute of Engineers and IAENG, Hong Kong. He is a life member of ISTE, India.

Dr. Paramartha Dutta, is a Professor in Department of Computer & System Sciences in Visva-Bharati in West Bengal, India. He was born 1966, earned his Bachelor's and Master's in Statistics from the Indian Statistical Institute, Calcutta in the years 1988 and

1990 respectively. He afterwards completed his Master of Technology in Computer science from the same Institute in the year 1993 and Doctor of Philosophy in Engineering from the Bengal Engineering and Science University, Shibpur in 2005.

He has served in the capacity of research personnel in various projects funded by Govt. of India, which include DRDO, CSIR, Indian Statistical Institute, Calcutta, etc. Dr. Dutta is now a Professor in the Department of Computer and System Sciences of the Visva Bharati University, West Bengal, India. Prior to this, he served Kalyani Government Engineering College and College of Engineering in West Bengal as a full-time faculty member. Dr. Dutta remained associated as Visiting/Guest Faculty of several Universities/Institutes encompassing West Bengal University of Technology, Kalyani University, Tripura University, National Institute of Technology, Arunachal Pradesh to name some.

He has coauthored eight books and has also edited eleven books with leading publishing houses such as Springer, Elsevier, John Wiley, Taylor and Francis, IGI Global, etc., to his credit. He has published more than 200 papers in various journals and conference proceedings, both international and national as well as several book chapters in edited volumes of reputed International publishing houses like Elsevier, Springer-Verlag, CRC Press, John Wiley, IGI Global, to name a few. Dr. Dutta has guided seven scholars who already had been awarded their Ph.D. In addition, three of his scholars have submitted their Ph.D. theses. Presently, he is supervising six scholars registered for their Ph.D. programs.

Dr. Dutta is the coinventor of six published Indian Patents and one published International Patent. Dr. Dutta as investigator, successfully implemented projects funded by AICTE, DST of the Govt. of India.

Prof. Dutta has served/serves in the capacity of external member of Boards of Studies of relevant departments of various Universities like West Bengal University of Technology, Kalyani University, Tripura University, Assam University, Silchar to name a few. He served as an expert on several Selection Boards conducted by West Bengal Public Service Commission, Assam University, Silchar, National Institute of Technology, Arunachal Pradesh, Sambalpur University, etc. Dr. Dutta is a Life Fellow of the Optical Society of India (OSI), Institute of Electronics and Telecommunication Engineering (IETE), Institution of Engineers of India (IE(I)), Life Senior Member of Computer Society of India (CSI), Life Member of Indian Science Congress Association (ISCA), International Association for Computer Science and Information Technology (IACSIT), International Association of Engineers, Hong Kong (IEng), Indian Society for Technical Education (ISTE), Indian

Unit of Pattern Recognition and Artificial Intelligence (IUPRAI) – the Indian affiliate of the International Association for Pattern Recognition (IAPR), Senior Member of Association of Computing Machinery (ACM), IEEE Computer Society (USA), and IEEE Computational Intelligence Society, USA.

Contributors

Rudranath Banerjee
Research Scholar
Department of Computer
    Science and Engineering
National Institute of
Technology Nagaland
Nagaland, India

Asit Barman
Assistant Professor of
    Department CSE/IT
Siliguri Institute of Technology
Salbari, Sukna
Darjeeling,
    West Bengal, India

Dr. Sankhayan Choudhury
Professor of Department
    Computer Science and
    Engineering
University of Calcutta
Salt Lake,
    Kolkata, India

Sourav De
Associate Professor
Department of Computer
    Science and Engineering
Cooch Behar Government
Engineering College
Cooch Behar, West Bengal,
    India

Shouvik Dey
Assistant Professor

Department of Computer
    Science and Engineering
National Institute of
Technology Nagaland
Nagaland, India

Dr. Paramartha Dutta
Professor of Dept. of Computer
    and System Sciences
Visva-Bharati University
Santiniketan,
    West Bengal, India

Avishek Nandi
Department of Computer and
    System Sciences
Visva-Bharati University
Santiniketan, West Bengal,
    India

Md Nasir
Department of Computer and
    System Sciences
Visva-Bharati University
Santiniketan, West Bengal,
    India

Varshini Prakash
Research Assistant
VIT University
Vellore, Tamil Nadu, India

Dr. B. K. Tripathy
Dean, School of Information
    Technology Engineering
VIT University
Vellore, Tamil Nadu, India

CHAPTER 1

Recent Advancements in Automatic Sign Language Recognition (SLR)

Varshini Prakash and B.K. Tripathy

Automatic Sign Language recognition remains a relevant and active research field, with upcoming techniques outperforming the existing developments. Automatic Sign Language Recognition has a broad social impact and can help to eliminate the communication barrier for the hearing impaired. This chapter studies - in chronological order - the existing techniques used in the recent years for SLR such as using Deep Neural Networks, 3D CNN, embedding of a CNN into HMM, adaptive pooling and their challenges and limitations. We also propose to discuss improved approaches which curtail the setbacks and would present an overview of the recent advancements and future extension in this area of research.

Keywords: Sign Language Recognition (SLR), Convolutional Neural Networks (CNN), Hidden Markov Model (HMM).

1.1 INTRODUCTION

Sign Language communication uses visual gestures and signs used by deaf people universally as their first language. Deaf people use sign language, a language with gestures as an alternative to spoken language. These gestures are a combination of movements of face, hand and different parts of the body. This is considered by deaf people as their natural way of communication. While the common people face difficulty in learning learning and comprehending sign languages, it is difficult for deaf people to learn oral languages. Sign Language Recognition approaches have aimed at eliminating this communication gap with these techniques advancing and improving along with the improvement in technology.

A sign language generally comprises signs for both complete words and also letters so as to represent words that cannot be enacted using corresponding signs in that language, for instance, names. It is quicker to perform a sentence as a sequence of sign words rather than performing it using the signs for letters. There are many sign languages: some of the popular ones are American Sign Language, Indian Sign Language, French Sign Language, and Chinese Sign Language.

Sign language has two main components - finger-spelling (posture) and dynamic hand movement (gesture). It is expressed through a varied mixture of manual articulations such as hand-signs, body movements, and additionally, even facial expressions [28]. Therefore, it is still a highly complex and challenging task to recognize sign language with high accuracy.

1.1.1 Common Challenges Faced in SLR

The challenges faced in Sign Language Recognition are some of the challenges generally faced in most computer vision applications. Occlusion or obstruction of objects, presence of adverse noise or disturbances can prove to be difficult while segmenting the hand signs and facial expressions. Furthermore, motion blur can be caused by moving objects and a cluttered background can make this task of segmentation harder [1]. The next step would be gesture spotting, which aims to determine meaningful gestures from a pool of all the gestures, some of which may be irrelevant.

The next big challenge is the processing of continuous sign sentences, which are a sequence of sign words. Similar to the problem of co-articulation in speech recognition, it is important to determine the beginning of one sign and the end of another in order to find the transition of signs [4]. Individual signs need to be identified from this continuous sequence of signs.

Difficulty in collaboratively exploiting information from the course or trajectory of hand and body movements is another challenge. One of the main complexities is the expression of these trajectories through descriptors. Developing hand sign descriptors is a tricky process since this requires hand regions to be tracked from images or frames in a video, and these regions have to be segmented or extracted from every frame from their respective backgrounds, which might sometimes be too cluttered. Motion trajectory description requires key points to be tracked, followed by curve fitting. The varying movements of hand and body joints and occlusion from visual representations often makes it difficult to obtain satisfying results.

The three stages in the standard approach to SLR are:

1. location of regions of importance or meaningful signs

2. feature extraction for the development of descriptors to describe these regions

3. classification is performed using the features obtained from the second stage

1.1.2 Approach

The two main approaches that have been used to recognize sign languages are:

1. Sensory gloves that use computer vision to detect joint movements [9]

2. Bayesian Belief networks or Neural Network classifiers that use sensing elements

The following are some of the **prominent models** that have been used for Sign Language Recognition in recent years:

Linear classifiers are simple to use and take less time to train and can achieve relatively high accuracies for the task of gesture and posture detection. However, complex non-linear models like Hidden Markov models (HMM), and Bayesian Networks are exceptions. The downside to these approaches is that they often require heavy pre-processing due to their experimental constraints and need for hand-crafted feature extraction techniques. The advantage Neural Networks have over these approaches is that they can automatically learn the key features required for classification. Although some neural networks like Feed Forward Neural networks do demand hand-crafted feature extraction.

Convolutional Neural Networks (CNNs) have proved to be efficient in perceiving gestures and events [36]. They do, however, possess other flaws. Although CNNs are applicable as filters, they often fail to retain all the spatiotemporal attributes of an image, which are vital for most gesture recognition tasks.

3D CNN was proposed to overcome the failure of CNNs to preserve spatial-temporal features [14]. 3D CNN is an extension of 2D CNN with an additional dimension in time in order to account for the temporal characteristics. The main difference is that convolutions are only applied on the spatial dimensions in two-dimensional CNNS and the features are computed from these 2D feature maps. While in three-dimensional CNNs, where videos are taken as the input, the motion information in continuous frames of the video is taken into account, along with the spatial features.

While Hidden Markov Models (HMMs) have been used predominantly in the field of Automatic Speech Recognition (ASR), they are an uncommon approach in the field of Computer Vision. Deep Convolutional Neural Networks (DCNNs) are generally preferred for computer vision tasks [3]. A CNN with multiple hidden layers is embedded within a HMM framework, trained as a hybrid CNN-HMM in an end-to-end fashion. The outputs from the CNN are regarded as true Bayesian posteriors [18].

The use of adaptive pooling and capsule networks [12] in CNNs have resulted in a high increase in the accuracy, as high as 99%. Human-Robot Interaction (HRI) can use an extension of this technique as non-verbal communication. The advantage of using an adaptive layer pooling is that it enables the network to construct fixed-length matrices by training with variable sized images. Such networks aid in improving the scale invariance. The robustness of the model is increased since the images in the dataset undergo slight rotations, making the task challenging. While an adaptive pooling layer provides scale invariance, capsule layers make the system invariant to any transitions, thereby, making the framework more robust.

1.2 MODELS USED FOR SLR

1.2.1 Hidden Markov Model

Hidden Markov Models satisfy the Markov Property which is the memoryless property of a stochastic process. This property was coined after the mathematician Andrey Markov. A Markov model incorporates the Markov assumption, which states that the probability distribution of an observation at a given time t depends only on the state at time $t - 1$ which produced the observation q_t.

Markov Assumption: $P(q_t = a|q_1...q_{t-1}) = P(q_t = a|q_{t-1})$

HMM is a probabilistic model which determines the probability of a sequence of observable events using hidden states. The probability of a hidden state is determined exclusively by the probability of the prior hidden state. A directed graph of states and possible transitions between them determine the possible sequences; they are called *topology*. The edge of this graph represents the probability of moving from one state to other [16].

The HMM are characterized by three fundamental problems: estimation, evaluation and decoding. The estimation problem is concerned with adjusting $\lambda = (A, B)$ to maximize $P(O|\lambda)$ given an observation sequence O [35]. The objective of the evaluation problem is to compute the probability that the observed sequence was generated by the model $P(O|\lambda)$. The decoding problem aims to recover the state sequence given the observation sequence.

A forward backward algorithm is used to generate the probability $P(O|\lambda)$ where π denotes the initial state probabilities, a denotes the transition state probabilities and b denotes the output probabilities.

$$\alpha_1(i) = \pi_i b_i(O_1) \forall i \text{ if } i \in S_\text{I}, \ \pi_\text{i} = \frac{1}{n_\text{I}} \text{ else } \pi_\text{i} = 0$$

Calculating $\alpha()$ for all the states j along the time axis $t = 2, .., T$

$$\alpha_t(j) = [\textstyle\sum_i \alpha_{t-1}(i)a_{ij}]b_j(O_t)$$

Final Probability is given by

$$P(O|\lambda) = \sum_{i \in S_F} \alpha_T(i)$$

The hands were tracked in two demonstrations, one without gloves and one with a solid coloured glove. The consequent shape, orientation and trajectory of the hands were fed as inputs to the HMM for sign word recognitions. HMMs possess some inherent properties that make them ideal for gesture recognition systems, typically having served as a baseline for such tasks. Both the training and recognition phases do not require explicit word segmentation. To include context models in the training of HMM, several methods such as embedded training were initially used [35]. The speech recognition community used this technology to include context and language models at several levels. Viterbi algorithm is the most popular algorithm for predicting optimal hidden state sequences, and it uses maximum posterior probability (MAP) for pattern recognition and classification. It decodes the signed sequence to find the best likelihood for a given sequence [5].

Traditionally, HMMs were used to model a single observation sequence. Their use has now been extended to model dissimilarities over multiple sequences. Despite their good grasp on perceiving time series data, their use is often restricted in handling discrete hidden variables one at a time in simple state spaces. They remain unsuccessful in understanding the dependencies among multiple information channels on these grounds. While HMMs handle the interactions in observation space, Coupled Hidden Markov Models (CHMM), an extension of HMMs model the interactions in state space. Multiple HMMs, each corresponding to a data stream are assembled together to form a CHMM. They have an edge over HMMs since their initial conditions are more robust, training speed quicker and model likelihood superior. CHMM with multi sensor data framework outperforms individual SLR systems based on individual sensors. This framework for dynamic hand gesture recognition is formed using two tensors which provide varying depth information and frames per second. The CHMM architecture incorporate temporal coupling in the sub-systems to represent the asynchronous and temporal inter modal dependencies between multiple information channels [20].

Adaptive HMMs adaptively determine the hidden states rather than predefining them to a fixed value. The adaptive states are obtained using the varying hand shapes. The Euclidean distance between the current frame and the previous frame of the video is calculated to get a vector. A threshold is set to segment the video by variation of hand shapes. If the distance is smaller than the threshold, it is labelled as 0, else as 1. This forms a binary sequence. After pre-processing, the hidden state of the sign is the segment number of successive '1'. The mode of the segments on training samples is the final hidden state for a specific sign word [41].

Entropy based k-means clustering is used to evaluate the number of states in the HMM model [24]. Unlike spoken languages, sign language cannot simplify the left-right structural constraint of a HMM to define the basic phoneme of a word. To fix this issue, entropy based K-means clustering is used to obtain the unit length of each word (phoneme based). Maximum entropy principle and knee point of entropy criterion are used to determine the number of clusters. Moreover, this approach combines the artificial bee colony (ABC) algorithm with the Baum Welch algorithm using a data driven method and thereby governs the structure of HMM. The ABC algorithm is an evolutionary algorithm used for optimization purposes. The Baum-Welch algorithm optimizes the parameters of HMM. However, this only computes the local optimum. The HMM framework is learned by the ABC algorithm and updates the values of the transition probability matrix accordingly.

The properties of HMM allows the representation of doubly stochastic processes with signal segmentation allow their effective use for sign language recognition.

1.2.2 Gaussian Mixture Model

Probability density functions are often estimated by modelling them as Gaussians. Only two parameters are required to define a Gaussian model where μ is the mean of the distribution and σ is the standard deviation of the distribution.

Univariate Gaussian Distribution

$$N(x|\mu, \sigma) = \frac{1}{\sqrt{2\pi\sigma^2}} e^{-\frac{(x-\mu)^2}{2\sigma^2}}$$

Multivariate Gaussian Distribution

$$N(x|\mu, \Sigma) = \frac{1}{\sqrt{2\pi|\Sigma|}} \exp\{-\frac{1}{2}(x-\mu)^T\Sigma^{-1}(x-\mu)\}$$

The mixture of Gaussians results in an accurate distribution of data. They only require relatively few parameters for estimation and can be learned from a dataset that is relatively small. A Gaussian mixture with k Gaussians N_i, each with its own μ, σ and weight w_i specifying the relative importance of the distribution N_i can be defined as

$$M(x) = \sum_{i=1}^{k} w_i N_i(x)$$

Gaussian mixtures are used to estimate the underlying class probability distribution by automatically learning the complex object motion trajectories. Video trackers and sign language measurements gathered from gloves wired with sensors are instances of object trajectories. Principal Component Analysis (PCA) of these trajectories are built, and this representation for each class is segmented and then fitted into the GMM. The recognition system is made more robust and noise tolerant by making the training set diverse; this results in a complex Probability Density Function (PDF).

The Expectation Maximization (EM) algorithm is used to fit mixtures of Gaussians to the data. It is an estimation problem consisting of a two-step iterative process; namely the E-step and the M-step. The means, variances and mixing coefficients are initialized and the initial log likelihood is evaluated.

1. E-step: The responsibilities are evaluated given the current parameter estimates.

2. M-step: The parameters are re-estimated using the current responsibilities.

The log likelihood of the training data is defined as follows:

$$L(x_1, ..., x_n) = \sum_{j=1}^{n} \log_2(M(x_j))$$

The log likelihood is evaluated after every iteration and the iteration repeats until the convergence criteria is satisfied. The EM algorithm is convergent monotonically and is used to detect and find a local maximum. Data from the training set is pruned, merged and split using a model splitting process to automatically estimate the number of modes as twice the maximum sub trajectories in all the trajectories for the class. The product of mixing the weight of a mode and the number of samples determines the number of input samples required for effective estimation of mode parameters. The new trajectories are classified by computing the log likelihood, once the GMM has been trained. The trajectory is classified into the class that is depicted by the GMM with the maximal likelihood [21].

A multimodal framework combines both manual and non-manual signs. Since it is challenging to extract a good feature set from a multimodal system, a classifier is required to pick only certain class samples. HMM-GMM models have been popularly used to recognize gestures. HMM is used as a temporal classifier and models the sequential dependencies in data. The output density of each state is calculated by defining a GMM for every state of that model.

The HMM is then trained for each sign gesture sequence. The output probabilities are re-estimated using the Baum-Welch algorithm and the Viterbi algorithm is used to perform recognition. The HMM-GMM approach mandates a systematic approach to tune the parameters by varying both the number of hidden states as well as the Gaussian mixture components. The role played by facial expressions has been understated in the field of SLR. However, the complexity of feature interpretation and understanding in the case of facial expressions results in them being disregarded. [21] claims both facial expression and hand gestures are essential to deliver information about the gesture.

While dealing with a large vocabulary SLR, the main challenge for sign language recognition is the presence of a large search space. The proposed solution is a fuzzy decision tree with heterogeneous classifiers. A divide-and-conquer method is used to hierarchically classify sign language attributes using multiple classifiers at each stage [7]. Unsustainable candidates are eliminated using a one- or two-handed classifier using a GMM. A Finite State

Machine (FSM) based method is used as hand shape classifier. Finally, a Self-organizing Feature Maps/Hidden Markov Model SOFM/HMM classifier is used to tackle signer independent difficulties. Signer independent difficulties are the challenges posed by effective extraction of common features from different signers and model convergence difficulties.

A GMM can determine if a gesture is demonstrated using one or two hands. This is concluded by the GMM by observing the left hand, which stays motionless for most one-handed signs, resulting in a very stable data. Therefore, the most stable frame is obtained. In the case of a two-handed sign, all the frames that include the left hand, which is in motion are included as a part of the training data. A motionless left hand allows the GMM to conclude that it is a one-handed sign and the training data is extracted accordingly. Training data only uses the position and orientation information of left hand for classification. Once the candidate words are classified under their corresponding one- or two-handed classes, these are used by the hand shape classifier.

1.2.3 Neural Networks

ANNs are inspired by the biological neuron. CNNs are a variant of ANN, influenced by the functioning of the human brain's visual cortex. The neurons in a CNN are modeled based on the biological nerve cells, connecting the receptive field, which are local regions of the visual field. Discrete convolutions are performed on the image with filter values as trainable weights to accomplish this connection. Feature maps are formed as a result of the application of multiple filters on each channel combined with the activation functions of the neurons [3].

One of the biggest advances is the use of Neural Networks in place of Gaussian Mixture Models (GMMs) [34, 27, 10]. GMMs are typically used to provide a representation of the relationship between HMM state and input feature. Any required level of accuracy can be achieved by modeling the probability distributions if the GMM has enough of a number of mixtures. However, GMMs pose some serious drawbacks. Firstly, they have proven to be statistically inefficient for modelling data that lies around a non-linear region in the data space [10]. Secondly, the number of parameters could drastically increase even if the feature size increases by a small amount. This could result in a performance degradation at higher dimensions. Neural Networks can provide a solution for both the above challenges posed by GMMs [22].

Vision-based analysis for Sign Language Recognition captures signs from a video camera. The camera captures signs acted out with a coloured glove. The system acquires the images, pre-processes the images, extracts features and finally performs gesture recognition using Recurrent Neural Networks (RNN). RNNs provide considerable advantage over feedforward networks especially for applications that demand temporal processing [26].

RNNs are a class of Artificial Neural Network ANN in which the connection between the nodes form a directed graph along the temporal sequence. RNNs have feedback connections within the other layers of the network and itself. This enables the network a characteristic of local memory, which in turn allows the network to store patterns and sequences and present this to the network more than once. The input pattern is forwarded along the network while the recurrent activations are propagated backwards to the context layer.

The two famous architectures for recurrent networks are: Elman and Jordan networks.

The Elman network is a three-layer network with context units. The hidden layer is a recurrent layer containing a recurrent link from all the nodes and itself in addition to standard feedforward connections. The Jordan network is akin to The Elman network in most aspects. The major exception is that while a feedback loop from the output layer to the hidden layer is not a part of the architecture in Elman Network, Jordan Network contains a feedback loop. This makes the network's behaviour stable and suitable for gesture recognition. The processing nodes use sigmoid as the activation function.

1.2.4 Convolutional Neural Network: Hidden Markov Model

Hidden Markov Models have been a tool for standard pattern recognition since they can infer sequences of hidden states from time-varying signals. Although HMM dominates the field of automatic speech recognition, it remains rather unpopular in the field of Computer Vision. This can be attributed to its poor image modelling capabilities by GMMs, which are generally used in frameworks that use HMM, as compared to CNNs.

This chapter focuses on an approach which integrates CNNs in the HMM framework [18].

- While embedding a deep CNN into HMM, the output of the CNNs are treated as true Bayesian posteriors.

- Hidden states of sign words are used as underlying targets to train the CNN in a top down approach.

- The hybrid CNN-HMM system is trained end to end, by optimizing the weights by just considering the video input and the gloss output.

The output in most CNN approaches is evaluated based on the correlation with the ground truth, often ignoring the temporal structures or domains. Frame level labels are often a prerequisite for CNNs. It is difficult to annotate datasets with frame level labels when dealing with real time footage, video representations or real-life datasets.

Video input is given as a sequence of images for which the model learns an unobserved sequence of words that best fit the corresponding sign. This sequence is found using the Bayes theorem, which finds the class with maximum posterior probability. The number of hidden states in HMMs is established, which is then used to model the sign words.

A dynamic programming-based tracking approach is applied to the images as a part of preprocessing, and later input to the CNN. Image preprocessing is used to track the right hand across a set of frames since the right hand plays a dominant role in signing. The distortion suffered by the video is compensated by enlarging the crop size. The images are processed pixel-wise and the average of all the images in the training data is subtracted.

The video is input as a contiguous stream of frames or images $x_1^T = x_1, ..., x_T$. Automatic SLR tries to learn and model x_1^T sequence of images which best fit w_1^N unknown sign words. Sign words are assumed to occur in a monotonous fashion, unlike translation to spoken language from sign language, where rearrangements are necessary. Best fit sequence is found using Bayes Decision rule and the objective is to maximize the posterior probability $P(w_1^N|x_1^T)$: $x_1^T \rightarrow [w_1^N]_{\text{opt}} = \arg\max_{w_1^N}\{P(w_1^N|x_1^T)\}$

The true class posterior probability when modelled by generative models is decomposed into a combination of two knowledge sources – a product of language model and visual model. The problem is modelled using HMM, a stochastic FSA to account for the temporal variation of the input. A first order Markov assumption is made and the Viterbi algorithm is used to maximize the posterior probability using the following equation:

$$[w_1^N]_{\text{opt}} = \arg\max_{w_1^N}\{p(w_1^N) \max_{s_1^T}\{\prod p(x_1^T|s_t, w_1^N)\, p(s_t|s_{t-1}, w_1^N)\}$$

The output probability of the HMM $p(x_1^T|s_t, w_1^N)$ is modelled by the CNN which have been known to model images better than generative models like HMM. The CNN models the posterior probability $p(s|x)$ since it is a discriminative model. CNN is used in a hybrid approach to model the posterior probability for a hidden state s given an input x.

A pooled state transition model that defines the transitions in the HMM in Bakis structure (left-to-right structure; forward, loops and skips across at most just one state, and two subsequent states share the same class probabilities) is applied across all sign-words $P(s_t|s_{t-1})$ is employed. The HMM models the garbage class as an ergodic state - a state that is aperiodic and positive-recurrent - with independent transition probabilities to make it more flexible, so as to make its insertion between sign words easier.

The architecture of the CNN employed incorporates three classifying layers. The network includes two intermediary auxiliary classifiers besides the final classifier. This encourages discrimination in lower stages of the network. The total loss is inclusive of the loss from these auxiliary classifiers. Each classifier is preceded by a dropout layer and all non-linearities are rectified linear units (ReLU).

A frame-state-alignment is to be obtained at the CNN training phase. A training and validation set is generated using this alignment to evaluate the accuracy per frame and stop the training at a good point, generally before the last few iterations. The model that results in the maximum accuracy on the automatic validation set is chosen once the CNN is trained. All three classifiers are used to estimate the iteration that performs best. The hybrid CNN-HMM includes a normalized exponential function whose resulting posteriors are used in the HMM as observation probabilities. In the tandem CNN-HMM approach the activations from the last layer before the softmax that yields the highest accuracy on the validation data are employed. Features are extracted from both train and test datasets for the Tandem system because a HMM-GMM system will be retrained with these features.

The HMM is based on the freely available state-of-the-art open source speech recognition system RASR [10]. The system performance is measured in Word Error Rate (WER), which is based on the Levenshtein alignment which computes the desired number of insertion, deletion and substitutions between reference and hypothesis sentence which results in a transformation of the hypothesis into the reference sequence.

$$\text{WER} = \frac{\#\text{deletions} + \#\text{insertions} + \#\text{substitutions}}{\#\text{referenceobservations}}$$

1.2.5   3D Convolutional Neural Network

CNNs are a popular choice for many computer vision applications such as object recognition, facial recognition, content-based image retrieval, motion tracking, optical flow, scene reconstruction, image restoration and so on [23]. CNNs can automate the process of feature construction. However, it struggles in the task of video classification due to its inability to incorporate both spatial and temporal information.

A 3D CNN model extracts both spatial and temporal features by performing 3D convolutions, thereby capturing the motion information encoded in multiple adjacent frames. Multiple channels of information are generated by the model from the input frames, and the final feature representation is obtained by combining information from all channels [14].

The main purpose of 3D CNNs is to obtain spatial and temporal attributes from video stream. Additionally, 3D CNNs can also be used to record motion information, without the aid of hand-crafted features that outline specific sign language movements - typically used in most existing methods. A 3D convolution is an extension of 2D convolution with an additional time dimension. The resulting 3D convolution provides the foundation for the 3D CNN architecture [11]. Both spatial and temporal features can be extracted by 3D CNNs from video stream. They can also capture motion information from raw data, unlike the use of hand-crafted features to describe sign language motion in the existing methods. A 2D convolution is extended to 3D convolution with added time dimension. This 3D convolution is used to construct the architecture for 3D CNNs [11].

A 3D CNN integrates hand shapes, motion trajectory and facial expressions. The dataset of sign words is composed of color, depth and body skeleton images [2]. The dataset is rendered from Microsoft Kinect, a motion sensing device which provides a colour and depth stream. Kinect is used as a capture device and extends the possibility of obtaining body joint locations in real time.

Hand shapes are discriminated using the changes in colour and depth in pixel levels. The additional time dimension is used to detect the motion trajectory using the variation of body joints. Since multiple visual sources are fed as input to CNN, it can learn the change in colour, depth and motion trajectory. The architecture of 3D CNN performs convolution of the integrated input and performs subsampling on adjacent video frames. Since GMM-HMM is a standard temporal classification technique, it has been regarded as the baseline. Motion trajectory attributes and hand-shape features are extracted to train the GMM-HMM. The algorithm includes both color and depth information. Experimental results show 3D CNNs outperforming GMM-HMM baselines.

To effectively capture and record the motion information from consecutive frames, 3D convolution is performed in the convolutional layers of CNN. Multiple continuous frames are stacked together as a cube and a 3D kernel is applied to this cube to achieve the 3D convolution.

$$\text{3D Convolution: } f_{xyt} = \tanh(\sum_{i,j,k} w_{ijk} v_{(x+1)(y+j)(t+k)} + b)$$

A 3D kernel can extract only a single kind of characteristic feature from the cube, since the kernel weights are reproduced across the entire cube. Multiple features are generated from the corresponding lower-level feature maps by increasing the number of feature maps in later layers. This is achieved by the application of multiple 3D convolutions to the same location in the previous layer with distinct kernels. The kernel weight and bias parameters are trained using supervised or unsupervised approaches.

Multiple convolution and sub-sampling layers are stacked together alternatively to construct a CNN architecture. The architecture comprises eight layers including the input layer. After the input layer, the first four layers after the input layer are convolution layers immediately followed by sub-sampling layer and two pairs of convolution and sub-sampling layers. Two fully-connected layers precede the output layer. Sub-sampling causes the model to be resistant against small spatial noise or distortions. Therefore, the subsequent layers perform pattern recognition with lower resolution at progressively larger spatial scales.

Video stream classification has been achieved using 3D CNNs [13, 14, 17]. The main concern in using 3D CNNs to handle video data is the time-consuming training process. However, the use of CUDA for parallel processing could help achieve real time efficiency. The developed 3D CNN model when trained using supervised algorithm requires a large number of labeled samples. The number of labeled samples can be significantly reduced when such model is pre-trained using unsupervised algorithms.

1.2.6 Restricted Boltzmann Machine

Being an energy-based model, it is used as a generative model in various applications to estimate probability or data distribution. RBM is composed of two kinds of layers which are independent of each other. The two layers comprise visible and hidden units. Although they are independent, the units of the other layer govern them. The Gradient-based Contrastive Divergence algorithm is used to train the RBM. The Gibbs sampling method is used in RBM to obtain a fitting estimate of the log likelihood gradient. Convergence of the model depends on proper adjustments made on RBM parameters, such as the learning rate, weights, number of hidden units, momentum and so on. RBM with fewer parameters can be be a good alternative in cases where CNN fails to generalize well.

RBM - a generative stochastic deep learning model - can learn a probability distribution over its inputs. This method uses a deep learning approach to perform automatic sign language recognition over visual data. RBM as a deep generative model is evaluated based on its ability to generate the input data probability distribution. The generative capabilities of the network provide an added advantage of fewer input data requirement and fewer parameters to achieve better generalization. Furthermore, the performance of the model can be enhanced by fusing different RBM blocks each of which correspond to a different visual modality.

This approach [32] uses two different input modalities, namely RGB and depth to enhance the accuracy of sign language recognition. Hands are detected from three kinds of input images by the CNNs: original, cropped and noisy cropped images. To improve the recognition accuracy, a CNN model built on Faster-RCNN is fine-tuned and then used to detect hand signs from the input image. The modalities of all three kinds of images are fed to the RBM, the output is then combined with an additional RBM in order to identify the hand sign label.

The model is made more robust and tolerant to noise and missing data by augmenting additional data from cropped images and noisy cropped images. Impulse noise or Salt and Pepper noise and Gaussian noise is added to the input data. The noise robustness of the model is evaluated against various kinds of noise and against the effect of hyper parameter tuning.

The modalities of all three different forms are fed to an RBM and these outputs are combined in another RBM to recognize the hand sign label. This process on the whole involves the use of six RBMs whose inputs are described below:

- RBM_1: Five RGB noisy cropped image.

- RBM_2: Five RGB cropped images.

- RBM_3: Hand detected from the RGB input image.

- RBM_4: Five noisy cropped images.

- RBM_5: Five depth cropped images.

- RBM_6: Original depth detected hand.

Different test cases are generated to compute the model's noise robustness, by incorporating various types of noise into the input. Output class labels are completely or partially generated using the input image labels and some states.

1.2.7 Adaptive Pooling and Capsule Network

Deep Learning Approaches commonly used for SLR are Convolutional Networks. This chapter also explores more recent techniques such as Adaptive Pooling and Capsule Networks. [12] proposes a posture learning framework for SLR which is robust towards image rotation and image quality.

This framework applies both the concepts of capsule networks and pooling layers simultaneously, confirming that the combined use of both the method enhances the speed of convergence and accuracy of the network. The architecture begins with three convolutions. The output is subsequently processed using two adaptive pooling layers and a 6D convolutional capsule layer. This intermediate output is fed into a fully connected capsule layer.

Adaptive pooling layers provide the benefit of allowing the network to be trained with images of varying size while at the same time producing matrices of fixed length. In-variance to scale can be enhanced by training the network with images of multiple sizes. The feature matrices produced by the adaptive pooling layer are fed to the capsule layer. The capsule layer makes the framework robust by providing transitional invariance. The images in the dataset are rotated slightly, making it quite challenging.

Adaptive Pooling - Pooling has played a key role in CNNs. Pooling or alternatively subsampling, allows the reduction of the spatial size of the representation. The added advantages are reduced parameters and computation and therefore, decreasing the chances of the network overfitting. However, some of the spatial features are lost during the pooling operation. [39, 19].

In spatial pooling, spatial features are preserved in the local spatial bins. The size is equivalent to the size of the input image. Max pooling method is used to store the output of the previous representation in each of the spatial bins. If there are n filters in the previous convolutional layer m spatial bins, the output will be a vector of dimension nxm.

Adaptive spatial pooling has proven to be effective for tasks that require high accuracy and tasks hat want to retain most of the spatial properties of an image. For instance, problems like object detection and classification have such requirements [25, 37]. Moreover, it provides the network with the ability to be trained with images of multiple sizes.

Capsule Network - Convolutional networks have connections in between layers through neurons. Capsule networks are connected through capsules. For each capsule (vector representation), the output of the capsule j can be denoted as:

$$v_j = g(s_j)$$

where v_j is the output vector of capsule j, s_j is the input to the capsule j and $g(.)$ is a squashing function which squashes the input into a small interval given by:

$$g(x) = \frac{||x||^2}{1 + ||x||^2} \frac{x}{||x||}$$

with x is the input vector of the function.

The sigmoid function squashes the vector between a small interval of 0 and 1. Capsule length is a representation of the probability that the correspondent entity or sub-entity exists. Each capsule consists of parameters that denote a variety of characteristics such as orientation, positional and scale-related information of a specific entity [33].

The weights in convolution kernels and transformation filters of the capsule network are updated using the backpropagation algorithm. The weights for coupling coefficient c and prior log probabilities b are updated using the routing phase [6]. The feature extractions that results from the vector transformation could be more robust than the scale transformation in a CNN. [33].

1.3   CURRENT APPLICATIONS

This section discusses state-of-the-art applications, identifies their limitations and potential scope to address these limitations.

1.3.1   Hand Gesture Recognition System

The construction of an automatic human action recognition system that uses spatiotemporal data can be broken down into two phases - feature extraction and classification [30]. The first phase involves capturing the features from consecutive frames. This results in descriptors. Descriptors are representations consisting of one or more feature vectors which helps the computer distinguish between different actions. The second phase deals with classifying the action. A classifier can discriminate individual signs or actions using these representations [29].

CNNs are used to automate the task of feature extraction while ANNs are used for classification. However, occlusion and motion clutter remains a major challenge for most of these applications. Restricting the model to a fixed frame of reference greatly limits its applicability. Some multiple viewpoint based action models have managed to find a solution by not fixating on a single and familiar viewpoint. The designing and application of such a product would be beneficial in public places like airports, hotels, railways and counters of banks, which are generally a challenge for those with a hearing disability. Moreover, if the recognized gestures are converted to speech, this can be further applied to aid speaking impaired people in delivery of speech or lectures.

1.3.2 Sensory Gloves

Gloves embedded with sensors work better than a camera in most cases since the user has an added advantage moving around freely and flexibly provided it is bound by a radius - the length of the wire that connects the glove to the computer. However, a camera does not provide this edge, restricting the user to remain in position in front of the camera. The advantage is even greater in the case of wireless gloves where the restriction can be further eliminated by using wireless gloves along with a wireless camera. The performance of the glove is not affected by light, electric field, magnetic field or any other interference.

Previously, the most common application for sensor gloves was to create virtual reality 3D environments for gaming. The gloves or a combination of these gloves along with other sensors were used as a input device to capture gestures or actions. The actions of the user wearing a glove are recorded and mirrored in the game to induce a sense of reality. These gloves have also been used to control robots using various commands or gestures.

Consecutive hand gestures are detected and then identified as a specific action which is used to perform a certain function, for instance as a command to control a robot. ANN with feed forward and back propagation algorithms have been used for this task of recognition. The movements and shape of other parts of the body excluding the hand such as face, legs, arms cannot be captured using these gloves. A major limitation of this application is that it is the only posture that can be considered, not the complete picture of the body. This method ignores the signs for letters such as j and z since it requires gestures that demand movement.

1.4 CONCLUSION AND FUTURE SCOPE

The presence of a Sign Language Recognition System can help eliminate communication barriers. The designing and application of such a product would be beneficial in public places like airports, hotels, railways and counters of banks, which are generally a challenge for those with a hearing disability. Moreover, this method can be used by an aphonic or speaking impaired person to deliver a speech. The main challenge in developing a real-time sign language recognition system with a high performance and adaptability is the requirement for the generation of a large dataset and the scalability of the network weights [38].

The existing techniques used for SLR in the recent years such as Hidden Markov Models, Gaussian Mixture Models, Neural Networks, Variants of Neural Network such as CNN, RNN, 3D CNN, RBM and some ensemble techniques such as CNN-HMM have been extensively studied. Some of the most advanced techniques that were studied are Neural Networks that enforce adaptive pooling and capsule networks. The complexity of the entire RBM entity or ensemble can be reduced by the definition of simple independent RBMs that can aid in splitting the information in the early training phase. Besides, the model can be extended to deal with a sequence of images from which it can model the spatiotemporal features of hand-signs.

In order to corroborate the viability of identifying a complex sequence of gestures automatically in a continuous manner, real-time Sign Language Recognition is necessary. Naturally, the priority for a low error rate high. Sign Languages are most often directly translated word by word in general, without considering the rules of grammar when it comes to sentence formation, as followed in spoken languages. While a sign language is being translated, it is must to heed to the rules of the intended spoken language. Addition of a speech engine to speak the translated text would help enhance ease of use.

Sensor gloves can only capture the hand gestures, they fail in capturing dynamic hand movements or facial expressions, which play a crucial role in facial recognition. This model can be extended with a technique to record and recognize facial expressions and hand movements. The realization of a combination of such actions requires multiple networks from different streams, that capture information from different parts of the body. Not much has been explored in the field of multiple viewpoints for action recognition in the recent years. Although, its training complexity is relatively high and there is scope for optimization [8]. Most of the work in sign language recognition is concerned with the actions of a single individual. Consequently, tracking the actions of multiple people in a video remains an unpopular problem [15].

Future extension of the approach that uses adaptive pooling and capsule networks work would be to upgrade the framework to non-static gestures. This could be achieved using recurrent neural network layers with capsules. The applications of SLR systems, both vision-based and glove-based have also been discussed.

GLOSSARY

Hidden Markov Model (HMM): A statistical Markov model in which the system being modeled is assumed to be a Markov process with hidden states.

Gaussian Mixture Model (GMM): A probabilistic model with the assumption that every data point is generated from a mixture of finite Gaussian distributions with unknown parameters.

Convolutional Neural Networks (CNN): A shift or space invariant neural network often used in computer vision attributing to their characteristics of weight sharing and translational invariance.

Capsule Networks: A neural network model that aims to mimic hierarchical relationships by forming a stable representation using the organization and reuse of structures called "capsules" in a CNN.

Adaptive Pooling: The pooling layers that are allow the network to be adaptable to images of varying size.

Restricted Boltzmann Machine (RBM): A generative stochastic deep learning model that can learn a probability distribution over its inputs.

FURTHER READING

Yang, Z., Shi, Z., Shen, X. and Tai, Y.W., 2019. SF-Net: Structured Feature Network for Continuous Sign Language Recognition.

Pu, J., Zhou, W. and Li, H., 2019. Iterative alignment network for continuous sign language recognition.

BIBLIOGRAPHY

[1] George Awad, Junwei Han, and Alistair Sutherland. A unified system for segmentation and tracking of face and hands in sign language recognition. In *18th International Conference on Pattern Recognition (ICPR'06)*, volume 1, pages 239–242. IEEE, 2006.

[2] Faisal Bashir, Ashfaq Khokhar, and Dan Schonfeld. Automatic object trajectory-based motion recognition using gaussian mixture models. In *2005 IEEE International Conference on Multimedia and Expo*, pages 1532–1535. IEEE, 2005.

[3] Vivek Bheda and Dianna Radpour. Using deep convolutional networks for gesture recognition in American sign language. *arXiv preprint arXiv:1710.06836*, 2017.

[4] MK Bhuyan, D Ghosh, and PK Bora. Continuous hand gesture segmentation and co-articulation detection. In *Computer Vision, Graphics and Image Processing*, pages 564–575. Springer, 2006.

[5] Terry Caelli and Brendan McCane. Components analysis of hidden Markov models in computer vision. In *ICIAP*, volume 3, page 510, 2003.

[6] Ruilong Chen, Md Asif Jalal, Lyudmila Mihaylova, and Roger K Moore. Learning capsules for vehicle logo recognition. In *2018 21st International Conference on Information Fusion (FUSION)*, pages 565–572. IEEE, 2018.

[7] Gaolin Fang, Wen Gao, and Debin Zhao. Large vocabulary sign language recognition based on fuzzy decision trees. *IEEE Transactions on Systems, Man, and Cybernetics-Part A: Systems and Humans*, 34(3):305–314, 2004.

[8] Ali Farhadi, Mostafa Kamali Tabrizi, Ian Endres, and David Forsyth. A latent model of discriminative aspect. In *2009 IEEE 12th International Conference on Computer Vision*, pages 948–955. IEEE, 2009.

[9] Sidney S Fels and Geoffrey E Hinton. Glove-talk: A neural network interface between a data-glove and a speech synthesizer. *IEEE transactions on Neural Networks*, 4(1):2–8, 1993.

[10] Geoffrey Hinton, Li Deng, Dong Yu, George Dahl, Abdelrahman Mohamed, Navdeep Jaitly, Andrew Senior, Vincent Vanhoucke, Patrick Nguyen, Brian Kingsbury, et al. Deep neural networks for acoustic modeling in speech recognition. *IEEE Signal processing magazine*, 29, 2012.

[11] Jie Huang, Wengang Zhou, Houqiang Li, and Weiping Li. Sign language recognition using 3d convolutional neural networks. In *2015 IEEE international conference on multimedia and expo (ICME)*, pages 1–6. IEEE, 2015.

[12] Md Asif Jalal, Ruilong Chen, Roger K Moore, and Lyudmila Mihaylova. American sign language posture understanding with deep neural networks. In *2018 21st International Conference on Information Fusion (FUSION)*, pages 573–579. IEEE, 2018.

[13] Hueihan Jhuang. *A biologically inspired system for action recognition*. PhD thesis, Massachusetts Institute of Technology, 2007.

[14] Shuiwang Ji, Wei Xu, Ming Yang, and Kai Yu. 3d convolutional neural networks for human action recognition. *IEEE transactions on pattern analysis and machine intelligence*, 35(1):221–231, 2013.

[15] Jaeyong Ju, Daehun Kim, Bonhwa Ku, David K Han, and Hanseok Ko. Online multi-person tracking with two-stage data association and online appearance model learning. *IET Computer Vision*, 11(1):87–95, 2016.

[16] Dan Jurafsky and James H Martin. *Speech and language processing*, volume 3. Pearson London, 2014.

[17] Andrej Karpathy, George Toderici, Sanketh Shetty, Thomas Leung, Rahul Sukthankar, and Li Fei-Fei. Large-scale video classification with convolutional neural networks. In *Proceedings of the IEEE conference on Computer Vision and Pattern Recognition*, pages 1725–1732, 2014.

[18] Oscar Koller, O Zargaran, Hermann Ney, and Richard Bowden. Deep sign: hybrid cnn-hmm for continuous sign language recognition. In *Proceedings of the British Machine Vision Conference 2016*, 2016.

[19] Alex Krizhevsky, Ilya Sutskever, and Geoffrey E Hinton. Imagenet classification with deep convolutional neural networks. In *Advances in neural information processing systems*, pages 1097–1105, 2012.

[20] Pradeep Kumar, Himaanshu Gauba, Partha Pratim Roy, and Debi Prosad Dogra. Coupled hmm-based multi-sensor data fusion for sign language recognition. *Pattern Recognition Letters*, 86:1–8, 2017.

[21] Pradeep Kumar, Partha Pratim Roy, and Debi Prosad Dogra. Independent bayesian classifier combination based sign language recognition using facial expression. *Information Sciences*, 428:30–48, 2018.

[22] Hai-Son Le, Ngoc-Quan Pham, and Duc-Dung Nguyen. Neural networks with hidden markov models in skeleton-based gesture recognition. In *Knowledge and Systems Engineering*, pages 299–311. Springer, 2015.

[23] Yann LeCun, Léon Bottou, Yoshua Bengio, Patrick Haffner, et al. Gradient-based learning applied to document recognition. *Proceedings of the IEEE*, 86(11):2278–2324, 1998.

[24] Tzuu-Hseng S Li, Min-Chi Kao, and Ping-Huan Kuo. Recognition system for home-service-related sign language using entropy-based *k*-means algorithm and abc-based hmm. *IEEE transactions on systems, man, and Cybernetics: systems*, 46(1):150–162, 2016.

[25] Yinglu Liu, Yan-Ming Zhang, Xu-Yao Zhang, and Cheng-Lin Liu. Adaptive spatial pooling for image classification. *Pattern Recognition*, 55:58–67, 2016.

[26] Manar Maraqa and Raed Abu-Zaiter. Recognition of arabic sign language (arsl) using recurrent neural networks. In *2008 First International Conference on the Applications of Digital Information and Web Technologies (ICADIWT)*, pages 478–481. IEEE, 2008.

[27] Abdel-rahman Mohamed, George E Dahl, and Geoffrey Hinton. Acoustic modeling using deep belief networks. *IEEE Transactions on Audio, Speech, and Language Processing*, 20(1):14–22, 2012.

[28] Tan Dat Nguyen and Surendra Ranganath. Facial expressions in american sign language: Tracking and recognition. *Pattern Recognition*, 45(5):1877–1891, 2012.

[29] Lionel Pigou, Sander Dieleman, Pieter-Jan Kindermans, and Benjamin Schrauwen. Sign language recognition using convolutional neural networks. In *European Conference on Computer Vision*, pages 572–578. Springer, 2014.

[30] Ronald Poppe. A survey on vision-based human action recognition. *Image and vision computing*, 28(6):976–990, 2010.

[31] Junfu Pu, Wengang Zhou, and Houqiang Li. Iterative alignment network for continuous sign language recognition. In *The IEEE Conference on Computer Vision and Pattern Recognition (CVPR)*, June 2019.

[32] Razieh Rastgoo, Kourosh Kiani, and Sergio Escalera. Multimodal deep hand sign language recognition in still images using restricted Boltzmann machine. *Entropy*, 20(11):809, 2018.

[33] Sara Sabour, Nicholas Frosst, and Geoffrey E Hinton. Dynamic routing between capsules. In *Advances in neural information processing systems*, pages 3856–3866, 2017.

[34] Frank Seide, Gang Li, and Dong Yu. Conversational speech transcription using context-dependent deep neural networks. In *Twelfth annual conference of the international speech communication association*, 2011.

[35] Thad Starner and Alex Pentland. Real-time American sign language recognition from video using hidden Markov models. In *Motion-Based Recognition*, pages 227–243. Springer, 1997.

[36] Vivienne Sze, Yu-Hsin Chen, Tien-Ju Yang, and Joel S Emer. Efficient processing of deep neural networks: A tutorial and survey. *Proceedings of the IEEE*, 105(12):2295–2329, 2017.

[37] Yi-Hsuan Tsai, Onur C Hamsici, and Ming-Hsuan Yang. Adaptive region pooling for object detection. In *Proceedings of the IEEE Conference on Computer Vision and Pattern Recognition*, pages 731–739, 2015.

[38] Di Wu, Nabin Sharma, and Michael Blumenstein. Recent advances in video-based human action recognition using deep learning: A review. In *2017 International Joint Conference on Neural Networks (IJCNN)*, pages 2865–2872. IEEE, 2017.

[39] Haibing Wu and Xiaodong Gu. Max-pooling dropout for regularization of convolutional neural networks. In *International Conference on Neural Information Processing*, pages 46–54. Springer, 2015.

[40] Zhaoyang Yang, Zhenmei Shi, Xiaoyong Shen, and Yu-Wing Tai. Sf-net: Structured feature network for continuous sign language recognition. *arXiv preprint arXiv:1908.01341*, 2019.

[41] Jihai Zhang, Wengang Zhou, Chao Xie, Junfu Pu, and Houqiang Li. Chinese sign language recognition with adaptive hmm. In *2016 IEEE International Conference on Multimedia and Expo (ICME)*, pages 1–6. IEEE, 2016.

CHAPTER 2

Distance-Shape-Texture Signature Trio for Facial Expression Recognition

Asit Barman, Sankhayan Chodhury, and Paramartha Dutta

I N this chapter we propose a combined distance-shape-texture signature trio feature for recognizing the expressions from face images. Distance, shape and texture signatures have great significance in recognizing emotions. The accurate landmarks extraction from the face image is a challenging task in the field of human emotion detection. An appearance model is used to identify the proper landmarks on the faces. These proper landmark points are used for the formation of a grid on the human face. Now triangles are identified among the grid and texture regions are marked from salient landmark points. The distance signature, shape signature and texture signature are determined from the grid and salient landmark points. The distance signature, shape signature and texture signature are used to get a distance-shape-texture signature trio feature for facial expression recognition.

Keywords: Distance Signature, Shape Signature, Texture Signature, Feature Selection, D-S-T signature trio for Facial Expression Recognition, MLP, RBF, NARX.

2.1   INTRODUCTION

In human society, facial expressions play an important role to express nonverbal communication. As the carrier of information, facial expressions are able to contain numerous information that can not be transferred by voice in the process of interpersonal communication. For that reason, facial expression recognition is a growing research topic in computer vision. Facial expression recognition has a great significant role in human-computer interaction technology and also has potential application value in many fields such as the medical field, computer-aided training and distance education. The traditional online tutoring system has the drawback of the interaction between students and teachers about teaching feedback. Emotional level based on facial expressions can solve the drawbacks of the normal tutoring system.

The facial expressions recognition system has three basic steps: image acquisition, facial feature extraction and classification. Ekman et al. [14] categorize the facial expressions into six categories such as happiness, sadness, anger, fear, surprise, and disgust. In order to encode the motion of various facial muscles, Ekman et al. developed the Facial Action Coding System (FACS) in which the facial movement is described by action units (AUs). Action unit identification has been used in many facial expression recognition systems.

The extensive survey of the facial expression recognitions are discussed in [24] and [15]. In [8], the frontal features are detected from images. These features are classified the expression into one among the seven classes. The Principal Component Analysis (PCA) is used to identify the action unit of asymmetric facial for facial expression recognition. The appearance based method active shape model [20] is used to identify the shape and texture features to recognize the expressions. The Candide nodes and Facial Action Units (FAUs) are used to find the geometrical displacement [19] for facial expression recognition. Authors select some points manually on the face image as a Candide grid node. The displacements of the Candide nodes are computed as the difference of the node between the first and the greatest facial expression intensity frame. These displacements were fed as an input to a Support Vector Machine (SVM) [13] system to classify the expressions.

The classification and recognition tasks are performed by Artificial Neural Networks. In this paper [17] the authors first consider the facial image. The eyes, mouth and eyebrows regions are considered to measure the distinctive human facial deformations as feature set. The discriminative features are fed as input of MLP to classify the facial expressions. The author [32] used geometric features and Gabor wavelet coefficients to categorize the emotions using multilayer perceptron. They showed that Gabor coefficients are giving promising results rather than geometric positions. The authors [9] used a single hidden-layer to recognize the facial expression of training patterns. The authors [16] used salient patch based features to classify the expressions.

In this chapter the Distance-Shape-Texture signature trio is used to identify human facial expressions. The distance signature, shape signature and texture signature are considered individually to form a signature trio. Effective landmark detection is a vital task in expression recognition. The Active Appearance Model (AAM) [12] is considered to identify the landmark points on a face. The few landmarks are extracted around the eye, nose, eyebrow and mouth region to constitute a grid. A grid is used to find the relative distances and normalized to get a distance signature. The triangles are formed within the grid. Normalized shape signatures are derived from triangles. The texture regions are extracted from salient landmark points and subsequently a Local Binary Pattern (LBP) [23] is also used to extract the texture features. These features are normalized to get the texture signature. These signature are combined to form a Distance-Shape-Texture signature trio for the recognition of facial expressions. The respective stability indices are obtained from the distance-shape-texture signature trio. The range, raw moment, entropy, kurtosis and skewness are also calculated from the distance-shape-texture signature trio. These extracted features are supplied to NARX, MLP and RBF networks as input to correctly recognize the expressions. The four benchmark datasets such as CK+, MUG, MMI and JAFFE are used to measure the effectiveness of the proposed system.

The rest of this chapter consists of the following sections. The proposed system is described in section 2.2. The facial landmark and subsequent signatures are elaborated on in section 2.3. The formation of Distance-Shape-Texture signature trio is described in section 2.4. The feature selection of Distance-Shape-Texture signature trio is described in section 2.5. The classification task is described in section 2.6. The experiments and results are shown in section 2.7. In the last section in 2.8 we draw the conclusion.

2.2 OVERVIEW OF THE PROPOSED SYSTEM

The schematic diagram is presented in Fig. 2.1 to show the proposed facial expression recognition. Active Appearance Model [29] is used to detect the landmark points on a face. The eyes, nose, mouth and eyebrows regions are discriminative due to the sensitivity of the emotions of the different face images. These points are considered in a way such that three 3-points on each eyebrow,

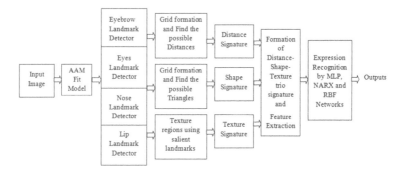

Figure 2.1: Proposed Block Diagram of Distance-Shape-Texture signature Trio

three points on nose, four points on mouth and four points on each eye as mentioned in Fig. 2.2. The detected landmark points form a grid and also try to compute the distances to get normalized distance signatures. Now triangles are identified from the grid node and corresponding normalized shape signatures are calculated. The texture regions are formed using detected landmarks. The Local Binary Pattern (LBP) is used on extracted regions to get the texture signature. Now individual normalized distance, shape and texture signatures are combined to form a distance-shape-texture signature trio. Stability indices of distance-shape-texture signature trio are determined from higher order of each distance, shape and texture signature. Statistical parameters such as range, raw moments, kurtosis, skewness and entropy are calculated from each distance-shape-texture signature trio to supplement the feature set. The extracted features are fed into MLP, NARX and RBF networks as input to classify the different expressions such as anger, fear, sadness, happiness, disgust and surprise separately.

2.3 FACIAL LANDMARK DETECTION

Proper landmark detection of facial expressions is a vital task in emotion recognition. Few facial landmarks deserve an important role due to their variation of expressions from one face to another one. There are many methods to extract the landmark points on the face images. In this chapter we consider a active appearance model (AAM) to extract the landmark points.

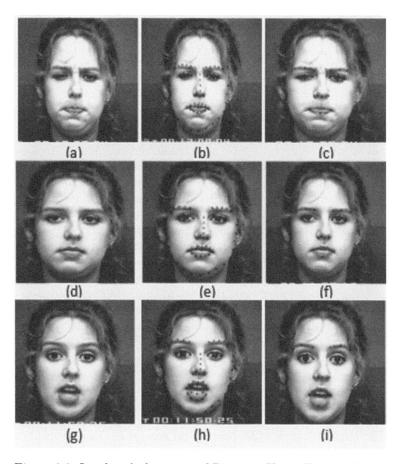

Figure 2.2: Landmark detection of Distance-Shape-Texture signature trio [(a, d & g) Original Image, (b, e & h) Landmarks detection using AAM and (c, f & i) Salient Landmarks]

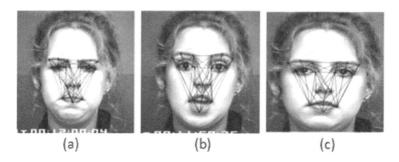

(a)   (b)   (c)

Figure 2.3: Formation of Grid and Triangle of Distance-Shape-Texture signature trio

Active Appearance Model: The accurate face alignment has a vital effect in a face recognition system. Active Appearance Model [29] is a well known method for appropriately locating objects. In the training phase of an active appearance model, huge face images with different shapes are considered in training. Then we mark a set of points to annotate face shape. The face shape is represented with the coordinate landmark points.

The transformation of the principal component analysis and the mean shape of all the faces are used to form a shape model for face alignment. Given a new face image, first we measure the initial position of model and then we calculate the suggested movements to get a good face alignment. The appearance parameters and shape parameters are obtained by the alignment for real time video tracking and pose estimate. The average speed of the appearance model is 123 frames/second. For this reason it is more applicable to image processing.

2.3.1   Grid Formation

The landmark points [28] play a vital role for the displacement of the facial components. In this context the salient landmark points are marked around nose, mouth, eyebrows and eyes regions due to their deformations of facial expression. However, the four points are marked on each eye, three points on eyebrow, four points on mouth and three points on nose. These selected landmarks are linked to form a grid [5] [3][4] [2] [6].

2.3.1.1   Distance Signature

The landmark points are extracted using AAM fitting model. The salient landmarks have the crucial role to represent the facial expression. The detected points are stored in a matrix mentioned as grid in Fig. 2.3. Each landmark is connected to the other to form a grid. Now we get the normalized distance as below

$$\gamma_{i1}^{f} = \frac{d_{g1}^{f}}{\sum_{g1=1}^{n1} d_{g1}^{f}}, \quad f = 1, 2, \ldots, m \text{ and } i1 = 1, 2, \ldots, n1 \quad (2.1)$$

where γ_{i1}^{f} represents the distance signature [5] [3] [2] [6] for image number f. Let d_{g1}^{f} be the distances. The distance and image index are represented by $i1$ and f. The total distances are $n1 = 210$.

2.3.2 Triangle Formation

The landmark points are extracted from the images using Active Appearance Model (AAM) [29] on the eyes, mouth, eyebrows and nose regions to identify the salient features. These landmarks are utilized for the formation of grid. The mid position of the nose landmark is marked as reference point to form the triangles for its central geometric location. We only consider those triangles holding the following properties.

Each triangle hold the following basic properties:

1. The sum of the angles in a triangle is 180 degrees. This is called the angle-sum property.

2. The sum of the lengths of any two sides of a triangle is greater than the length of the third side. Similarly, the difference between the lengths of any two sides of a triangle is less than the length of the third side.

3. The side opposite to the largest angle is the longest side of the triangle and the side opposite to the smallest angle is the shortest side of the triangle.

2.3.2.1 Shape Signature

These landmarks are used to form a grid. The triangles are identified within the grid. The perimeter is $p = a + b + c$, where a, b, c represent the edges. We also consider as $j = p/2$. The area is computed as $r = \sqrt{j(j-a)(j-b)(j-c)}$. The triangles are calculated to determine the shape as

$$s_i = \frac{r_i}{p_i^2} \quad (2.2)$$

Here the index of the triangles is specified by i. The normalized shape signature [3] [2] is computed as

$$\mu_i^f = \frac{s_i^f}{\sum_{i=1}^q s_i^f} \quad f = 1, 2, \ldots, m \text{ and } i = 1, 2, \ldots, g \qquad (2.3)$$

where g represents the total shape numbers, and m represents the total images. Here, μ_i^f represents the shape signature and s_i^f are the shapes of triangles as applicable.

2.3.3 Texture Region

The active facial patches play an effective role in facial expression [4] [6]. The patches are selected on a face image due to the variation of facial regions such as eyes, nose, mouth and eyebrows. The landmark points are extracted using an appearance model. These landmark points extract the eye corner, lip corner, mouth corner, nose corner and brow corner region. In this context we consider the size of the patches as 9×9. The indexed patches depending on their active region are shown in figure 2.4. Salient active patches $T_1, T_2, T_3, T_4, T_5, T_6, T_7$ and T_8 are derived from nose, lip corner, eyebrows and eyes respectively as depicted in Fig. 2.4. T_9 and T_{10} represent the nose corner points. The lip corner points are represented by T_{11} and T_{12}. T_{15} and T_{16} are situated just below the lip corner and T_{17} is the position between T_{15} and T_{16} patches. Now T_{13} and T_{14} are located as the adjacent region of the nose corner points. Now these extracted facial patches are combined to get a texture region of a particular facial image.

The process of texture region implies the following sequence of events:

1. Identification of the landmark points using AAM.

2. Prominent landmarks are marked on nose, eyebrows, eyes and mouth regions.

3. These salient landmarks help to extract the texture regions from facial image .

2.3.4 Local Binary Pattern

Local Binary Pattern [23] is considered for robustness to overcome the illumination. LBP calculates a binary number by using neighboring points with respect to centering points values. The pattern

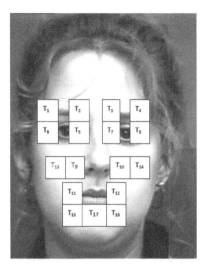

Figure 2.4: Position of facial patches of Distance-Shape-Texture signature trio

of pth bit sample is computed as

$$LBP(x,y) = \sum_{p=0}^{j-1} s(g_p - g_c)2^p \qquad (2.4)$$

where g_c indicates the centering pixel value, g_p represents the pixel values of neighborhood at coordinate points of (x,y) and j represents the bit sample of total number.

$$s(x) = 1\, if\, x \geq 1 = 0, x \leq 0 \qquad (2.5)$$

The histogram of the LBP is defined as

$$h_u = \sum_{xy} LBP(x,y) = u, \quad u = 0,1,\ldots,l-1 \qquad (2.6)$$

The different labels are represented by l. In this experiment a uniform pattern is used to extract features. The bit-wise transition from 1 to 0 and 0 to 1 vice-versa of a circular bit sample is used for uniform measure. It is showed that when we consider a texture image then it is proved that above 90% information is kept in the (8,2) neighborhood.

2.3.4.1 Texture Signature

The landmark points are extracted by an appearance based model. Now salient landmarks are considered to identify the active facial patches of the facial expression. Salient landmarks are considered to extract the active facial patches. These active facial patches are merged together to get a texture region. Now LBP is applied on that texture region to obtain a texture descriptor. A normalized texture signature [16] [6] is defined as below

$$\beta_{i2}^{f} = \frac{h_{g2}^{f}}{\sum_{g2=1}^{n2} h_{g2}^{f}}, \ f = 1, 2, \ldots, m \ \text{and} \ i2 = 1, 2, \ldots, n2 \quad (2.7)$$

where β_{i2}^{f} indicates the texture signature, f indicates the image number. Let h_{g2}^{f} represent the texture descriptors. Here $i2$ represents the texture descriptor. Here m represents the number of images. Hence $n2 = 59$ indicates the texture descriptors.

2.4 FORMATION OF DISTANCE-SHAPE-TEXTURE SIGNATURE TRIO FOR FEATURE EXTRACTION

We use distance, shape and texture signatures individually to form a Distance-Shape-Texture (DST) signature trio for facial expression recognition. This signature trio is not directly fed as input of different neural networks to identify the expressions. The stability index is computed from the DST trio as a vital feature to identify the expressions of different persons. The statistical measures of skewness, kurtosis, range, moment and entropy are computed from the DST trio to get a prominent feature set.

2.4.1 Stability Index of Distance-Shape-Texture signature trio

Stability index is a formidable feature in facial expression recognition. We already defined the definition of distance, shape and texture signatures using equations 2.1, 2.3, 2.7. Higher order signatures are calculated from three signatures. These higher signatures are considered to get stability index as

$$(\mu_{i1}^{r1})^{f} = \frac{(d_{g1}^{r1})^{f}}{(\sum_{g1=1}^{n1} d_{g1}^{r1})^{f}} \quad (2.8)$$

where the $r1^{th}$ signatures are obtained from each signature (distance signature, shape signature and texture signature). In our experiment we use the order as $r1 = 1, 2, \ldots, 10$ for distance signature $r1 = 1, 2, \ldots, 8$ for shape signature and $r1 = 1, 2, \ldots, 8$ for texture signature. Now the differences of the $r1^{th}$ order signatures are consider as

$$\Delta(d_{i1}^{r1})^{f} = (\mu_{i1}^{r1})^{f} - (\mu_{i1}^{r1-1})^{f} \quad (2.9)$$

Now a threshold is selected based on the analysis of variations of the higher signatures. This threshold follows the constraint $(\Delta d_{i1}^{r1})^f \leq \zeta_1$ for a small threshold. The tolerance threshold is fixed as $\zeta1$ after evaluation of the experimentation of distance, shape and texture signatures. Such orders $r1$ are defined as stability indices of each signature.

2.4.2 Statistical measures from Distance-Shape-Texture signature trio

Promising facial expression recognition is done by the involvement of more numbers of features. The statistical parameters such as skewness, moments and entropy are used to extract more numbers of features set from normalized signatures.

The moment [31] up to $4th$ order is defined as

$$P_k = \sum_{i=1}^{n} w(i,k) \times v(i) \tag{2.10}$$

where $v = (v_1, v_2, \ldots, v_n)$ represents the normalized signature vector of each signature and $w(i,k)$ is i raised to power of k, the range of the k is $k = 1, 2, 3, 4$. Skewness [31] is used to measure the asymmetry of the data. The skewness and kurtosis [11] are formulated as

$$Q = \frac{\sum_{i=1}^{n} w(i,3)^2 \times v(i)}{\sum_{i=1}^{n} w(i,2)^3 \times v(i)} \text{ and} \tag{2.11}$$

$$R = \frac{\sum_{i=1}^{n} w(i,4)}{\sum_{i=1}^{n} w(i,2)^2 \times v(i)} \tag{2.12}$$

The entropy [11] is also formulated from normalized distance, shape and texture signatures as

$$E1 = -\sum_{i1=1}^{n1} p_{i1} \log(p_{i1}) \tag{2.13}$$

where p_{i1} represents the relevant relative frequencies of the distance, shape and texture signatures and $\sum_{i1=1}^{n1} p_{i1} = 1$, $0 \leq p_{i1} \leq 1$.

2.5 FEATURE SELECTION OF DISTANCE-SHAPE-TEXTURE SIGNATURE TRIO

Each higher order signature is defined to compute the stability indices as per equations 2.8, 2.9. In the same way statistical parameters such as moment, entropy, skewness, range and kurtosis are evaluated as per equations 2.10, 2.11, 2.12 and 2.13. Now we combine all the features such as moment, skewness, range, kurtosis and entropy to form a DST trio feature set. Each of the DST trio signatures has a 27-dimensional vector feature. A 27-dimensional vector feature is fed to NARX, MLP and RBF networks as input for categorization of emotions.

2.6 CLASSIFICATION OF DISTANCE-SHAPE-TEXTURE SIGNATURE TRIO FEATURES

Three artificial networks (RBF, MLP and NARX) are used to classify the expressions on four benchmark datasets such as Cohn-Kanade, JAFFE, MMI and MUG.

2.6.1 Multilayer Perceptron

The distance-shape-texture signature trio is represented by a vector of 27 elements of each facial image. These features are also normalized and fed as input variable to the network of MLP [27]. It produces the output depending on the minimum error. The training process is completed by minimizing the mean square error [10] as $e_k = 1/2 \sum (t_k - y_k)^2$ for six classes. Here t_k and y_k represent the target and desired response. The different class labels set as $k = 6$ in the training procedure.

The four steps of the training process are mentioned as initialization of weights, feed-forward, minimizing the errors through back propagation and modification of biases and weights. In the first step, small values are assigned to initial weights. The input unit represents as x_i $(i = 1, \dots, n)$, where n indicates the inputs of received signal. After that it transmits an input signal to each hidden unit z_j $(j = 1, 2, \dots, 8)$. The biases are denoted as w_o and v_o. The $tanh$ activation function is used to compute the output of each hidden unit. These outputs are transmitted to each output unit z_j. A linear activation function is used to get the output response. During backpropagation, each output unit computes its y_k for comparing with target t_k to compute the desired error. Depending on these errors, the δ_k $(k = 1, \dots, m)$ factor is calculated as output y_k and sent into the back of all the previous layer. In this experiment m indicates the output (The m value is 6). In a similar way, δ_j $(j = 1, \dots, p)$ factor is considered for every hidden unit z_j. In this experiment we set the hidden units as $p = 8$. In the last phase biases and weights are modified using the factor of δ. The threshold and learning rate are denoted as ζ and η. The stopping training criterion is $e_k \leq \zeta$. In the testing phase, six basic expressions are identified depending on the maximum output values.

2.6.2 Training using Nonlinear Auto Regressive with exogenous input

The NARX [18] network is a architecture of dynamical neural network. The network receives as input a window of past input and output values and computes the current output. The NARX model is defined as

$$y(t) = f(u(t-Du), \ldots, u(t-1), u(t), y(t-Dy), \ldots, y(t-1)) \quad (2.14)$$

Here, $u(t)$ and $y(t)$ represent the inputs and outputs of the network at time t respectively. The input and output time lags are denoted by Du and Dy, and f represents a nonlinear function which is approximated by backpropagation through time [18]. The backpropagation through time network has three layers as input, hidden and output layers. The output layers have the recurrent connections from the hidden layers. The hidden layer computes a linear or nonlinear transformation function of the input unit based on the activation function. The outputs $y(t)$ and hidden states $h(t)$ at time t are derived as

$$h(t) = \phi(w_{hx}x(t)) + b_h \text{ and } y(t) = \gamma(w_{yh}h(t)) + b_y \quad (2.15)$$

where the weight of input to hidden and hidden to output matrices are represented as w_{hx} and w_{yh}. The biases are denoted as b_h and b_y. The hyperbolic tangent functions (ϕ) use in hidden unit. In the same way the linear transfer function (γ) is used for output unit. In the training phase, the NARX network optimizes to get optimal weights by minimizing the errors between the predictions y and the target outputs.

2.6.3 Radial Basis Network

The radial basis network [26] has the three layers such as input, hidden and output. Here n represents the input and m indicates the output. The hidden layer reside in the middle position of input and output layers. The hypothetical connections are formed from the the input layers to hidden layers. In the same way, the weighted connections are formed from the hidden layers to output layers. The Gaussian function computes the output of RBF. In the training phase, the training datasets are labeled into six classes. The D-S-T signature trio is computed from the same training set and fed as input of that network. During the testing, extracted features are considered to measure the system performance of expressions recognition. The output of the RBF is classified into six categories beyond the expressions.

Here, we propose the training mechanism of the RBF. In the first step, small random values are assigned as weighted to input units. Each input unit x_i ($i = 1, \ldots, n$); here n represents the input receiving feature. This input feature are transferred to each hidden unit and calculates the radial basis function. A radial function is chosen by the centers among the input features. In the hidden layer, a radial basis function is used as an activation function to transform the input vector: $v_i(x_i) = \exp(-\frac{1}{2\sigma_i^2}||x_{ji} - X_{ji}||^2)$. Here $||.||$ represents a norm of the input feature, X_{ji} represents the centers of the input features, σ_i indicates the ith width of input and x_{ji} indicates the variable of j_{th} input. Now we consider the weights for the output layers with small random values. Now it calculates the output as

$$y_{ne} = \sum_{i=1}^{h} w_{im} v_i(x_i + w_o) \qquad (2.16)$$

where hidden nodes h, y_{ne} represent the m_{th} output value in the output layer for the n_{th} in coming pattern, w_{im} indicates the weight between i_{th} unit and m_{th} output node and w_o indicates the bias at n_{th} output. The network computes the minimum error using mean square error to stop the training.

2.7 EXPERIMENT AND RESULT

The four benchmark datasets are considered to test the system performances. We assess the D-S-T signature trio based facial expression recognition by applying the Cohn-Kanade (CK+) [21], JAFFE [22], MMI [30] and MUG [1] databases. The D-S-T signature trio is expressed as 27 variables of feature vector. The MLP RBF and NARX networks are considered for classification of expressions. The result comparison of the different state-of-art methods are evaluated here to justify the performance of the proposed system. The six basic emotions are considered as anger (Ang), disgust (Dis), fear (Fea), happy (Hap), surprise (Sur) and sadness (Sad).

2.7.1 Experiment on CK+ Database

The Cohn-Kanade (CK+) database consists of 100 university students aged between 18 and 30 years. To evaluate the system performance, a total of 393 images/samples are selected to measure the system performances of three different networks. In this dataset we used 70% images for training, 15% images are testing and remaining 15% for the validation. The total images are considered as 43 images for fea, 72 images for ang, 84 images for hap, 54 images for dis, 71 images for sad and 69 images for sur. We used three

TABLE 2.1 The Confusion Matrix of D-S-T signature trio on CK+ Dataset using MLP

	Ang	Dis	Fea	Hap	Sad	Sur
Ang	72	0	0	0	0	0
Dis	2	52	0	0	0	0
Fea	1	0	41	0	0	0
Hap	0	4	0	80	0	0
Sad	0	0	0	0	71	0
Sur	0	0	0	0	0	69

TABLE 2.2 The Confusion Matrix of D-S-T signature trio on CK+ Database using NARX

	Ang	Dis	Fea	Hap	Sad	Sur
Ang	70	2	0	0	0	0
Dis	0	52	2	0	0	0
Fea	0	0	41	1	1	0
Hap	0	0	0	83	1	0
Sad	0	0	0	0	69	2
Sur	0	0	0	0	0	67

artificial networks such as RBF, NARX and MLP to categorize the expressions into different categories.

Table 2.1 represents the performance of expressions of the trio signature using MLP. Anger, sadness and surprise are classified correctly. Disgust recognizes 52 samples properly and 2 samples are misclassified with anger. Happiness identifies 80 images properly and 4 images are misidentified with disgust. Fear identifies 41 images and 1 image is confused anger.

The NARX network is evaluated to measure the system performances of different expressions of the trio signature which is available in Table 2.2. Anger classifies properly 70 images and 2 samples are misclassified with disgust. Disgust recognizes 52 images properly and 2 samples are misclassified with anger. Fear classifies 41 images are properly and 2 samples are confused with sadness and surprise. Happiness recognizes 83 images and 1 image is confused as sadness. Sadness identifies 69 samples properly and 2 images are wrongly classified as surprise. In the same way 67 images are classified correctly as surprise.

The RBF network is also evaluated to test the system performances of the trio signature which is available in Table 2.3. From this table we see that anger and sadness are classified properly. But fear identifies correctly 41 samples but 2 samples are misclassified as surprise. Happiness recognizes properly 82 images and 2 samples are misidentified as surprise and anger. Surprise classifies

TABLE 2.3   The Confusion Matrix of D-S-T
signature trio on CK+ Database using RBF

	Ang	Dis	Fea	Hap	Sad	Sur
Ang	70	0	0	0	0	0
Dis	0	54	0	0	0	0
Fea	0	0	41	0	1	1
Hap	1	0	0	82	0	1
Sad	0	0	0	0	71	0
Sur	0	1	0	0	1	67

67 samples correctly and 2 samples are misclassified as sadness and disgust.

2.7.2 Experiment on JAFFE Dataset

The JAFFE [22] dataset has a total of 213 gray level images/samples of emotions. These expressions are six common and one neutral of the Japanese database. In this experiment we use a total of 97 images to measure the performances. The dataset is divided into 70% for training, 15% for testing and 15% validation.

Table 2.4 indicates the performance of recognized expressions of the trio signature using MLP. Anger, disgust, happiness and surprise are identified correctly. Fear recognizes correctly 11 samples but 3 samples are misidentified with anger, surprise and sadness. Sadness identifies 13 samples properly and 1 sample is confused as fear.

TABLE 2.4   The Confusion Matrix of D-S-T
signature trio on JAFFE Database using
MLP

	Ang	Dis	Fea	Hap	Sad	Sur
Ang	21	0	0	0	0	0
Dis	0	20	0	0	0	0
Fea	1	0	11	0	1	1
Hap	0	0	0	14	0	0
Sad	0	0	1	0	13	0
Sur	0	0	0	0	0	14

The NARX network is used to measure the system performances of different expressions of the trio signature which is available in Table 2.5. Anger identifies correctly 20 images and 1 image is misclassified with disgust. Disgust recognizes 18 samples rightly and 2 samples are misidentified as happiness. Fear identifies 13 samples correctly and 1 sample is confused as anger. Happiness recognizes 12 samples correctly and 2 samples are misclassified as sadness. Sadness identifies correctly 12 images and 2 images are wrongly classified with surprise. In the same way the 11 images are classified correctly as surprise and 1 image is confused with fear. The experimentation result of the RBF network is tabulated

TABLE 2.5 The Confusion Matrix of D-S-T signature trio on JAFFE Database using NARX

	Ang	Dis	Fea	Hap	Sad	Sur
Ang	20	1	0	0	0	0
Dis	0	18	2	0	0	0
Fea	1	0	13	0	0	0
Hap	0	0	0	12	2	0
Sad	0	0	0	0	12	2
Sur	0	0	1	0	0	11

in Table 2.6 to measure the performance of different expressions. From this table we observe that anger, disgust, happiness and surprise are classified correctly. But fear identifies 13 samples rightly but 1 sample is confused as disgust. Sadness identifies 13 images properly and 1 image is misclassified with fear.

TABLE 2.6 The Confusion Matrix of D-S-T signature trio on JAFFE Database using RBF

	Ang	Dis	Fea	Hap	Sad	Sur
Ang	21	0	0	0	0	0
Dis	0	20	0	0	0	0
Fea	0	1	13	0	0	0
Hap	0	0	0	14	0	0
Sad	0	0	1	0	13	0
Sur	0	0	0	0	0	14

2.7.3 Experiment on MMI Database

The MMI database is a more challenging dataset due to their spontaneous expressions. In this experiment we use a total of 113 images/samples to measure the performance of expressions. They were evaluated as 70% for training, 15% for testing and the remaining 15% for validation.

Table 2.7 indicates the performance of identified expressions of the trio signature using MLP. Anger identifies correctly 28 images and 12 images are misclassified with all expressions except anger. Disgust recognizes properly 30 images and 9 images are

TABLE 2.7 The Confusion Matrix of D-S-T signature trio on MMI Database using MLP

	Ang	Dis	Fea	Hap	Sad	Sur
Ang	28	2	4	0	5	1
Dis	3	30	3	1	1	1
Fea	1	1	16	1	5	1
Hap	1	3	4	12	1	3
Sad	1	0	0	1	27	1
Sur	0	1	1	3	1	15

TABLE 2.8 The Confusion Matrix of D-S-T signature trio on MMI Database using NARX

	Ang	Dis	Fea	Hap	Sad	Sur
Ang	37	3	0	0	0	0
Dis	4	33	2	0	0	0
Fea	0	0	23	1	1	0
Hap	0	0	0	23	1	0
Sad	0	0	0	3	26	1
Sur	0	1	0	0	0	18

confused with anger, fear, happiness, sadness and surprise. Fear recognizes 16 images properly and 9 images are misidentified with the remaining expressions. Happiness identifies 12 samples rightly, but 12 samples are misclassified with remaining expressions. Sadness recognizes 27 samples properly and 3 samples are misclassified as anger, happiness and surprise. Surprise recognizes 15 samples rightly and 6 samples are misclassified as happiness, sadness, disgust and fear.

The NARX network is also used to measure the performances of different expressions of the trio signature which is available in Table 2.8. Anger classifies correctly 37 images and 3 images are misidentified as disgust. Disgust recognizes properly 33 images and 6 images are wrongly classified with anger and fear. Fear classifies 23 samples are rightly and 2 samples are misclassified as happiness and sadness. Happiness recognizes 23 images properly and 1 image is confused as sadness. Sadness identifies 26 samples rightly and 4 samples are misclassified as happiness and surprise. In the same way 18 images are classified correctly as surprise and 1 image is confused with disgust.

The experimentation result of RBF network is also evaluated in Table 2.9 to show the performance of different expressions of the trio signature. Anger recognizes properly 34 images and 6 images are misidentified as disgust and happiness. Disgust identifies correctly 37 samples and 2 samples are misidentified as fear and surprise. Fear classifies properly 19 images and 6 images are

TABLE 2.9 The Confusion Matrix of D-S-T signature trio on MMI Database using RBF

	Ang	Dis	Fea	Hap	Sad	Sur
Ang	34	3	1	2	0	0
Dis	0	37	1	0	0	1
Fea	1	2	19	0	2	1
Hap	2	1	1	14	3	3
Sad	1	4	2	3	18	2
Sur	2	0	1	1	0	17

TABLE 2.10 The Confusion Matrix of D-S-T signature trio on MUG Database using MLP

	Ang	Dis	Fea	Hap	Sad	Sur
Ang	17	0	0	0	1	1
Dis	1	29	1	0	0	0
Fea	1	1	12	0	0	0
Hap	0	3	1	26	0	0
Sad	1	0	0	0	19	1
Sur	0	0	1	0	0	25

misidentified as anger, disgust, sadness and surprise. Happiness recognizes 14 images properly and 10 images are misclassified with remaining expressions. Sadness recognizes 18 images perfectly and 12 samples are misidentified with remaining expressions. Surprise identifies 17 samples correctly and 5 samples are misidentified as fear, happiness and surprise.

2.7.4 Experiment on MUG Database

MUG [1] dataset is evaluated to measure the performances of expressions. In this experiment we select a total of 197 images to classify the expressions. The dataset is divided into 70% for training, 15% for testing and 15% validation.

Table 2.10 represents the effectiveness of facial expressions of the trio signature using MLP. Anger classifies correctly 17 images and 2 images are misclassified as sadness. Disgust recognizes properly 29 images and 2 images are misidentified as anger and fear. Fear recognizes 12 samples perfectly and 2 samples are misclassified as anger and fear. Happiness identifies 26 samples rightly and 4 samples are confused as disgust and fear. Sadness identifies 19 samples are correctly and 2 samples are misclassified anger and surprise. Surprise correctly classifies 25 images and 1 image is confused as fear.

The NARX network is used to measure the performances of expressions of trio signature which is available in Table 2.11. Anger

TABLE 2.11   The Confusion Matrix of D-S-T signature trio on MUG Database using NARX

	Ang	Dis	Fea	Hap	Sad	Sur
Ang	18	1	0	0	0	0
Dis	0	29	2	0	0	0
Fea	0	0	13	1	0	0
Hap	0	0	0	29	1	0
Sad	0	0	0	0	19	2
Sur	0	1	0	0	0	23

TABLE 2.12   The Confusion Matrix of D-S-T signature trio on MUG Database using RBF

	Ang	**Dis**	**Fea**	**Hap**	**Sad**	**Sur**
Ang	19	0	0	0	0	0
Dis	0	31	0	0	3	1
Fea	0	0	14	0	0	0
Hap	2	1	1	26	0	0
Sad	0	0	1	0	20	0
Sur	0	1	0	0	0	25

classifies 18 images properly and 1 image is misidentified as disgust. Disgust recognizes 29 images properly and 2 samples are misidentified with fear. Fear recognizes 13 samples perfectly and 1 sample is misclassified as happiness. Happiness recognizes 29 images and 1 image is confused with sadness. Sadness identifies 19 images perfectly and 2 images are wrongly classified as surprise. In the same way 23 images are classified correctly as surprise and 1 image is misclassified with disgust.

The experimentation result of RBF network is tabulated in Table 2.12 to show the effectiveness of facial expressions of the trio signature. From this table we observe that anger and fear are classified correctly. Disgust identifies 31 images properly and 1 image is misidentified as surprise. Happiness identifies correctly 26 images and 4 images are confused as anger, disgust and fear. Sadness recognizes properly 20 images and 1 image is misidentified as fear. Surprise recognizes 25 images are properly and 1 image is wrongly identified with disgust.

TABLE 2.13 Testing Recognition rate on publicly available Databases of Six basic expressions with Different State-of-the-Art methods

	Ang	Dis	Fea	Hap	Sad	Sur	Avg.
CK+ Dataset							
DST (MLP)	100	96.3	95.3	95.2	100	100	97.8
DST (NARX)	97.2	96.3	95.3	98.8	97.2	100	97.4
DST (RBF)	97.2	100	95.3	97.6	100	97.1	97.8
[5]	96.1	95.7	95.8	98.4	95.9	95.8	96.3
[3]	98.1	100	95.8	100	93.8	100	97.9
[4]	98.6	98.1	97.7	98.8	98.6	100	98.6
[2]	100	100	100	100	100	100	100
[6]	98.6	100	97.7	94	98.6	98.6	97.9
[16]	87.8	93.3	94.3	94.2	96.4	98.4	94.1
[25]	87	91.5	90.9	96.9	84.5	91.2	90.3
[33]	76.2	94.1	86.1	96.3	88.2	98.7	91.5
JAFFE Dataset							
DST (MLP)	100	100	78.6	100	92.9	100	95.2
DST (NARX)	95.2	90	92.9	85.7	85.7	91.7	90.2
DST (RBF)	100	100	92.9	100	92.9	100	97.6
[5]	100	100	78.6	85.7	100	100	94
[3]	100	95.2	71.4	92.8	92.8	100	92.0
[4]	85.7	90.0	85.7	92.9	85.7	100	90.0
[2]	100	100	78.6	100	100	100	96.4
[6]	100	100	100	100	71.4	100	95.2
[16]	100	86.2	93.7	96.7	77.4	96.6	91.7
[25]	89.3	90.7	91.1	92.6	90.2	92.3	91.1
MMI Dataset							
DST (MLP)	70.0	76.9	64.0	50.0	90.0	71.4	70.3
DST (NARX)	92.5	84.6	92.0	95.8	86.7	94.7	91.1
DST (RBF)	85.0	94.9	76.0	58.3	60.0	81.0	75.9
[5]	82.8	81	71.4	87.5	100	66.7	81.5
[3]	100	85.7	64.2	100	61.1	80	81.8
[4]	97.5	94.9	96	91.7	93.3	100	95.5
[2]	100	85.7	64.3	100	61.1	80	81.9
[6]	77.5	97.4	92	87.5	90	86.4	88.5
[25]	80.1	78.2	81.3	83.2	77.1	81	80.1
[33]	65.6	72.5	72.5	88.2	71.1	93.8	77.4
MUG Dataset							
DST (MLP)	89.5	93.5	85.7	86.7	90.5	96.2	90.3
DST (NARX)	94.7	93.5	92.9	96.7	90.5	95.8	94.1
DST (RBF)	100	100	100	86.7	95.2	96.2	96.4
[5]	93.5	100	100	95.6	96.6	94.4	96.7
[3]	96.7	100	100	100	93.1	100	98.3
[4]	94.7	96.8	92.9	93.3	95.2	100	95.7
[2]	96.8	100	96.2	100	93.1	100	97.7
[6]	100	96.8	100	96.7	100	100	98.6

2.7.5 Comparison Analysis with Three Artificial Networks and State-of-the-Arts

The comparison task of the recognition rate of expressions of CK+ dataset with neural networks are presented in Table 2.13. We also show the comparison of expression recognition rate with different state-of-the-art. The MLP based DST trio signature achieves 97.8% average recognition rate with 100% recognition rate for anger, disgust and surprise and 95.2% as lowest recognition rate for happiness. The NARX based DST trio signature acquires 97.4% average recognition rate with correct recognition for surprise and 95.3% as lowest recognition for fear. In the same way, the RBF based DST trio signature achieves 97.8% average recognition with correct recognition for disgust and sadness and 95.3% as the lowest recognition rate for fear. The proposed MLP, NARX and RBF based DST trio signature methods are considered to measure the comparison with the existing literatures [5] [3][4] [2] [6] [33] [16] [25] [34]. It is also established that proposed MLP and NARX based procedures give promising recognition rates for expressions.

The JAFFE dataset is also considered for the comparison of the recognition rate of different neural networks which is shown in Table 2.13. The MLP based DST trio signature achieves 95.2% average recognition rate with 100% recognition rate for disgust, anger, happiness and surprise and 78.6% as the lowest recognition rate for fear. The NARX based DST trio signature acquires 90.2% average recognition rate with the highest recognition for anger and 85.7% as the lowest recognition for happiness. In the same way, the RBF based DST trio signature achieves 97.6% average recognition rate with correct recognition for anger, disgust, happiness and surprise and 92.9% as the lowest recognition rate for fear. The proposed MLP, NARX and RBF based DST trio signature methods are considered to measure the comparison with the existing literatures [5] [3][4] [2] [6] [33] [16] [34]. It is also established that proposed RBF based procedure gives promising recognition rates for expressions.

The MMI dataset is also evaluated for comparison task of recognition rate of expressions of different neural networks. Table 2.13 shows the comparison of facial expression recognition rates with different state-of-the-art. The MLP based DST trio signature achieves 70.3% average recognition rate with a 90% recognition rate for sadness and 50% as the lowest recognition rate for happiness. The NARX based DST trio signature acquires 91.1% average recognition rate with correct recognition for happiness and 84.6% as the lowest recognition for disgust. In the same way, the RBF based DST trio signature achieves 75.9% as the average recognition rate with the highest recognition rate for disgust and 60% as the lowest recognition rate for sadness. The proposed MLP, NARX and RBF based DST trio signature methods are considered to measure the comparison with the existing literatures [5] [3][4] [2] [6] [33] [16] [25]. It is also established that proposed NARX based procedure yields promising recognition rates for expressions.

The MUG dataset is also evaluated for the comparison task of recognition rates with neural networks which is available in Table 2.13. We also show the comparison of expression recognition rate with different state-of-the-art. The MLP based DST trio signature achieves a 90.3% average recognition rate with a 96.2% recognition rate for surprise and 85.7% as the lowest recognition rate for fear. The NARX based DST trio signature acquires a 94.1% average recognition rate with the highest recognition rate for happiness and 90.5% as the lowest recognition rate for sadness. In the same way, the RBF based DST trio signature achieves 96.4% as an average recognition rate with correct recognition for disgust and fear and 86.7% as the lowest recognition rate for happiness. The proposed MLP, NARX and RBF based DST trio signature methods are considered to measure the comparison with the existing literatures [5] [3][4] [2] [6]. It is also established that the proposed RBF based procedures yield more promising recognition rates than the MLP.

2.8 CONCLUSION

This chapter proposed a distance-shape-texture signature trio based facial expression recognition of three artificial neural networks such as MLP, NARX and RBF. Three different networks are used to evaluate the recognition rate of expressions. The performance of the proposed procedure is experimented by the recognition rate and comparison with the state-of-the-art. The experimental results also show the promising recognition rate of facial expressions of distance-shape-texture signature trio.

BIBLIOGRAPHY

[1] N. Aifanti, C. Papachristou, and A. Delopoulos. The mug facial expression database. In *in proc. 11th Int. Workshop on Image Analysis for Facial Expression Database*, pages 12–14, Desenzano, Italy, April 2010.

[2] Asit Barman and Paramartha Dutta. Facial expression recognition using distance and shape signature features. *Pattern Recognition Letters*, 2017.

[3] Asit Barman and Paramartha Dutta. Facial expression recognition using shape signature feature. In *Research in Computational Intelligence and Communication Networks (ICRCICN), 2017 Third International Conference on*, pages 174–179. IEEE, 2017.

[4] Asit Barman and Paramartha Dutta. Texture signature based facial expression recognition using narx. In *Calcutta Conference (CALCON), 2017 IEEE*, pages 6–10. IEEE, 2017.

[5] Asit Barman and Paramartha Dutta. Facial expression recognition using distance signature feature. In *Advanced Computational and Communication Paradigms*, pages 155–163. Springer, 2018.

[6] Asit Barman and Paramartha Dutta. Facial expression recognition using distance and texture signature relevant features. *Applied Soft Computing*, 77:88–105, 2019.

[7] Asit Barman and Paramartha Dutta. Influence of shape and texture features on facial expression recognition. *IET Image Processing*, 13(8):1349–1363, 2019.

[8] Marian Stewart Bartlett, Gwen Littlewort, Ian Fasel, and Javier R Movellan. Real time face detection and facial expression recognition: Development and applications to human computer interaction. In *Computer Vision and Pattern Recognition Workshop, 2003. CVPRW'03. Conference on*, volume 5, pages 53–53. IEEE, 2003.

[9] Hayet Boughrara, Mohamed Chtourou, Chokri Ben Amar, and Liming Chen. Facial expression recognition based on a mlp neural network using constructive training algorithm. *Multimedia Tools and Applications*, 75(2):709–731, 2016.

[10] Debasmita Chakrabarti and Debtanu Dutta. Facial expression recognition using eigenspaces. *Procedia Technology*, 10:755–761, 2013.

[11] CH Chen and David Stork. Handbook of pattern recognition & computer vision. *International Journal of Neural Systems*, 5(3):257, 1994.

[12] Timothy F Cootes, Gareth J Edwards, Christopher J Taylor, et al. Active appearance models. *IEEE Transactions on pattern analysis and machine intelligence*, 23(6):681–685, 2001.

[13] Corinna Cortes and Vladimir Vapnik. Support-vector networks. *Machine learning*, 20(3):273–297, 1995.

[14] P. Ekman and W. V. Frisen. *Emotion in the human Face*. NJ: Prentice Hall, Englewood Cliffs, 1975.

[15] Beat Fasel and Juergen Luettin. Automatic facial expression analysis: a survey. *Pattern recognition*, 36(1):259–275, 2003.

[16] SL Happy and Aurobinda Routray. Automatic facial expression recognition using features of salient facial patches. *Affective Computing, IEEE Transactions on*, 6(1):1–12, 2015.

[17] Ioannis Hatzilygeroudis Isidoros Perikos, Epaminondas Ziakopoulos. Recognizing emotions from facial expressions using neural network. In *IFIP International Conference on Artificial Intelligence Applications and Innovations*, pages 236–245. Springer, 2014.

[18] Herbert Jaeger. *Tutorial on training recurrent neural networks, covering BPPT, RTRL, EKF and the "echo state network" approach*, volume 5. GMD-Forschungszentrum Informationstechnik, 2002.

[19] Irene Kotsia and Ioannis Pitas. Facial expression recognition in image sequences using geometric deformation features and support vector machines. *Image Processing, IEEE Transactions on*, 16(1):172–187, 2007.

[20] Andreas Lanitis, Chris J Taylor, and Timothy F Cootes. Automatic interpretation and coding of face images using flexible models. *Pattern Analysis and Machine Intelligence, IEEE Transactions on*, 19(7):743–756, 1997.

[21] Patrick Lucey, Jeffrey F Cohn, Takeo Kanade, Jason Saragih, Zara Ambadar, and Iain Matthews. The extended cohn-kanade dataset (ck+): A complete dataset for action unit and emotion-specified expression. In *Computer Society Conference on Computer Vision and Pattern Recognition-Workshops*, pages 94–101. IEEE, 2010.

[22] Michael Lyons, Shota Akamatsu, Miyuki Kamachi, and Jiro Gyoba. Coding facial expressions with gabor wavelets. In *Automatic Face and Gesture Recognition, 1998. Proceedings. Third IEEE International Conference on*, pages 200–205. IEEE, 1998.

[23] Timo Ojala, Matti Pietikäinen, and David Harwood. A comparative study of texture measures with classification based on featured distributions. *Pattern recognition*, 29(1):51–59, 1996.

[24] Maja Pantic and Leon JM Rothkrantz. Automatic analysis of facial expressions: The state of the art. *Pattern Analysis and Machine Intelligence, IEEE Transactions on*, 22(12):1424–1445, 2000.

[25] Ahmad Poursaberi, Hossein Ahmadi Noubari, Marina Gavrilova, and Svetlana N Yanushkevich. Gauss–laguerre wavelet textural feature fusion with geometrical information for facial expression identification. *EURASIP Journal on Image and Video Processing*, 2012(1):1–13, 2012.

[26] Mark Rosenblum, Yaser Yacoob, and Larry S Davis. Human expression recognition from motion using a radial basis function network architecture. *Neural Networks, IEEE Transactions on*, 7(5):1121–1138, 1996.

[27] SN Sivanandam and SN Deepa. *Introduction to neural networks using Matlab 6.0*. Tata McGraw-Hill Education, 2006.

[28] Yun Tie and Ling Guan. Automatic landmark point detection and tracking for human facial expressions. *EURASIP Journal on Image and Video Processing*, 2013(1):8, 2013.

[29] Georgios Tzimiropoulos and Maja Pantic. Optimization problems for fast aam fitting in-the-wild. In *Proceedings of the IEEE international conference on computer vision*, pages 593–600, 2013.

[30] M. F. Valstar and M. Pantic. Induced disgust, happiness and surprise: an addition to the mmi facial expression database. In *Proceedings of Int'l Conf. Language Resources and Evaluation, Workshop on EMOTION*, pages 65–70, Malta, May 2010.

[31] Tzay Y Young. *Handbook of pattern recognition and image processing (vol. 2): computer vision*. Academic Press, Inc., 1994.

[32] Zhengyou Zhang. Feature-based facial expression recognition: Sensitivity analysis and experiments with a multilayer perceptron. *International Journal of Pattern Recognition and Artificial Intelligence*, 13(06):893–911, 1999.

[33] Lin Zhong, Qingshan Liu, Peng Yang, Junzhou Huang, and Dimitris N Metaxas. Learning multiscale active facial patches for expression analysis. *IEEE transactions on cybernetics*, 45(8):1499–1510, 2015.

[34] Lin Zhong, Qingshan Liu, Peng Yang, Bo Liu, Junzhou Huang, and Dimitris N Metaxas. Learning active facial patches for expression analysis. In *Computer Vision and Pattern Recognition (CVPR), 2012 IEEE Conference on*, pages 2562–2569. IEEE, 2012.

CHAPTER 3

Face Expression Recognition using Side Length Features Induced by Landmark Triangulation

Avishek Nandi, Paramartha Dutta,
and Md Nasir

The human face is capable of generating six basic types of facial expressions: Anger, Fear, Disgust, Sadness, Happiness, and Surprise. Each expression is reflected in the face as deformation and dislocation of facial components of a given face image. The Active Appearance Model (AAM) is a statistical model of object shape recognition used to generate landmark points describing the geometrical position of the facial components. Landmark points generated by AAM are used to formulate triangulation description of the shape of the face. In this article, the authors have proposed a novel method using a different ratio of side lengths of a triangle constituting geometrical shape derived through forming triangulation generated out of landmark points indicating axial positions of facial components fitted by an AAM. These novel side length features are learned by a MultiLayered Perceptron (MLP) network to segregate facial expression into six basic classes of expressions. The results obtained henceforth by applying the proposed method in JAFFE, Extended Cohn-Kanade (CK+), MUG and MMI databases are quite impressive.

3.1 INTRODUCTION

Affective analysis is one of the important aspects of Human-Computer Interaction(HCI) in the field of emotion analysis [1]. Automatic analysis of human emotions has many real-life applications in medical, computational and security and surveillance, monitoring systems, companion robotics, and mood-based suggestion systems. Some studies have found that among many different forms of human emotions the one emanating from the face has the highest power to differentiate one emotion from another. Thus, expression of the face contributes a significant amount of characteristic information for automated recognition of human emotions. To classify the different types of emotions, some sort of grouping of emotions in distinct classes is necessary. For this purpose, Ekman et al. proposed six classes of atomic emotions with a substantial difference in appearance and emotional quality: Anger(AN), Fear(FE), Disgust(DI), Sadness(SA), Happiness(HA), and Surprise(SU) [2]. Later they found that these six expression classes have almost the same arousal and variance characteristics among different sets of people from different geographical locations [3]. These six basic expressions are uniquely identifiable because a single class of emotion emulates a unique combination of muscular movement hence generating classifiable feature cues corresponding to that particular class of emotion. Another important consideration towards automatic facial expression recognition is that the quality of the detected feature plays an important role in the segregation of facial expression in different classes. The quality of facial feature descriptors is mainly hindered due to variation in camera angle, lighting conditions, low image resolution, low visibility conditions, head rotation, face occultation, etc. The primary challenge of the affective computing community is to design an efficient and effective affect descriptor that is independent of such conditions mentioned earlier. The design of an effective facial feature descriptor is an essential criterion for affect classification in a dynamically changing environment. Facial Action Coding System (FACS) is one of such systems which describes movement of facial muscles to classify each facial expression and effectively associate it to its respective expression class [4]. FACS is a powerful feature descriptor but Action Unit (AU) detectors of FACS suffer from misclassification in the case of occultation and pose variation. In contrast to FACS system, the geometric based methods directly use location and shape information of relevant facial components such as eyes, eyebrows, nose, and mouth to extract the emotion-related features from the face induced by location points which are fed into a classifier without any intermediate stage of Action Unit (AU) detection.

This paper is organized as follows, apart from the first introductory section, section 3.2 discusses the related works. Section 3.3 and section 3.4 include the gap analysis and motivation for the proposed work. The proposed methodology is described in section 3.5 along with a flow diagram of computation. The methodology section has four subsections. Subsection 3.5.1 deals with the detection of a facial component from the raw face image. Subsection 3.5.2 introduces the techniques we have used for the formation of triangles. Subsection 3.5.3 and 3.5.4 discuss the extraction of features and classification learning respectively. Section 3.6 contains detailed discussions and performance comparisons with experimental results on CK+, JAFFE, MMI, and MUG databases individually described in subsection 3.6.1 to 3.6.4 respectively. Section 3.7 concludes our study with the future scope of the study. Section 3.8 contains the acknowledgments.

3.2 RELATED WORKS

Facial recognition achieved a big success in Viola-Jones's works on face detection which in turn created a lot of research interest within the community in the field of affective computing in [5]. They have used the Haar-like feature to create an integral image to detect rectangular features. In contrast to that Sang et al. used a boosted-LBP based method for low-resolution images for efficient detection of facial expressions in [6], and the work by Wang and Wang introduces a self-quotient image-based filter to enhance the detection accuracy of Haar features of Viola Jones's algorithm in [7]. They have also used DCT with Gabor filters for the expression recognition task. In the work by Happy and Routray, they proposed a group of salient facial patches to detect and extract appearance cues in a pose and in an illumination independent manner [8]. The contemporary work of Llbeygi and Hosseini introduces a fuzzy inference system with genetic algorithm-based parameter tuning to recognize the emotion of the color face images in an efficient manner [9]. The facial expression recognition machine developed by Gomathi et al. uses an adaptive neuro-fuzzy learning (ANFIS) network for learning LBP features from a partitioned facial image [10]. In the pioneering work of Barman and Dutta, they have used a distance-shape signature of the geometry of a face-image in conjunction with a stability measure for enhancing classification accuracy in a pose invariant manner in [1]. They have extended their foundational work on distance and shape signature features with texture and shape co-relevant features, shape only features, distance only features and distance-texture features to improve the accuracy of expression recognition in an efficient manner [12, 13, 14, 15, 16]. In the paper [17] Turan et al. used a Soft Locality Preserving Map-based method to deal with high dimensional feature set for expression learning. Dhavalikar and Kulkarni suggested an AAM based landmarks combined with skin color texture to attain a good classification accuracy [18]. Liu et al. used a Deep Belief network for classification of shape feature of a static image which worked very well in different variable conditions of [19]. Some other works on Deep Convolutional Neural Network (DCNN) for static facial expression recognition can be found in [20, 21].

3.3 GAP ANALYSIS

The works we have discussed earlier suffer from person independent learning of facial features. The texture-based methods in [6, 7] suffer from variation in positions of facial components. The geometric based methods in [8, 1] lose the texture information but it is free from rotational influence. The deep learning-based methods presented in [19, 20, 21] suffer from extensive hardware requirements for learning the facial features. To overcome these limitations we have used a powerful geometric based feature with MultiLayer Perceptron(MLP) as a classification learner.

3.4 MOTIVATION

Pose, individual shape and illumination defying expression detection are a challenging problem in the field of self-regulated emotion recognition of the face. The expressional changes on the face always overlap with the original shape of the face and differentiating the affective changes from the original shape of the face is a challenging task. The selection of a proper shape descriptor more sensitive towards the affect relevant shape characteristics is an important criterion for a shape predictors model. We have introduced a novel shape extractor model to overcome those previously stated issues which calculates side lengths of a descriptor triangle covering the individual components of eyes, eyebrows, nose, mouth and outer region of the face to construct a combined shape signature feature.

The descriptor triangles are fitted on the face to identify the deformation on the face caused by muscular tension.

3.5 PROPOSED METHODOLOGY

The flow chart in figure 3.1 shows the flow of the computation of the proposed machine.

3.5.1 Facial Component Detection

The human face consists of four important facial components: eye, eyebrow, nose, and mouth. The change in expression in the face is reflected in the form of variance in the shape and structural changes of these basic components. Hence the accurate detection of shape of those components plays an important role in the computation of expression relevant features. We have used the Active Appearance Model (AAM) which is a texture-based facial component detection model for extraction of the geometrical position of individual components generally termed as landmarks. The model we have used for fitting the landmark points is AAM FAST-SIC by Tzimiropoulos et al. [2] AAM FAST-SIC shape predictor computes the axial position of components by appearance analysis of the face, which is in turn displayed through plotting the coordinate points on the eyebrows, eyes, nose and lips and perimeter region of the face. The model we have used plots 68 landmark points on the face which are mainly on corner and border regions of facial components. These 68 points can be subdivided as follows: 12 points at outer lips border, 8 points on inner lips border, 11 points on

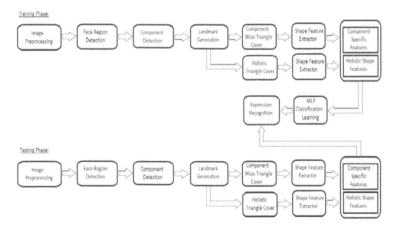

Figure 3.1: Flow chart illustration of the proposed method.

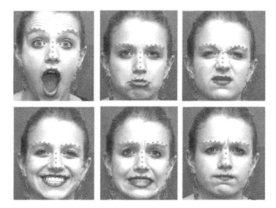

Figure 3.2: Landmarks induced by AAM FAST-SIC [2] plotted over candidate images showing geometric positions of landmark points on the face.

the nose, 6 points on each of the eyes, 5 points on each of the eyebrows, and 17 points on the outer region of the face. These induced landmarks are shown in figure 3.2.

3.5.2 Formation of Triangles

The geometrical shape of each facial component is associated with the structure of the area covered by relevant landmark points. To capture the shape-based data of any geometric object formation a trigonometrical shape formation is necessary. A triangle is the most basic form shape which we have used in our proposed method. The formation of triangles involves three landmark points which are selected from a set of landmark points for each component, for the sake of simplicity. Initially, all possible combinations from landmark points are calculated. Then the points from the set are used to form triangles. The six basic combination sets are computed for two eyes, two eyebrows, nose, and mouth. Another combination set is constructed by taking a reduced set of landmarks from the set of 68 AAM generated points as salient landmark points from each component of the face, hence creating a holistic purview of the face. Figure 3.3 shows the salient landmark points covering all the regions of the face.

These six combination sets of left eye, right eye, left eyebrow, right eyebrow, nose, mouth, and holistic face are used to form triangles covering the individual components and also for the unified face. The formation of component-wise triangulation and face triangulation is displayed in Figures 3.4 and 3.5 respectively.

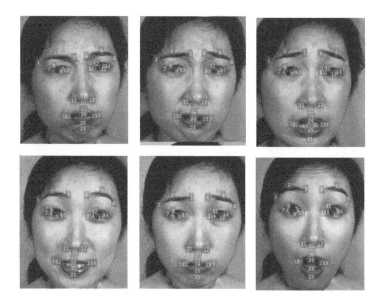

Figure 3.3: Selected salient landmark points plotted over the face.

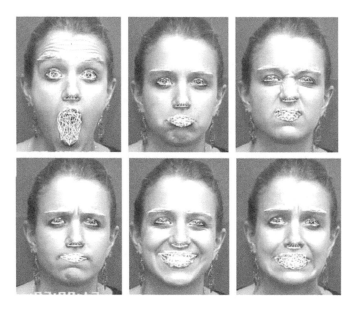

Figure 3.4: Triangle formation for each component on the face.

3.5.3 Feature Extraction

Extraction of proper features plays an important role in terms of accuracy of detection of expression of the face in different variable

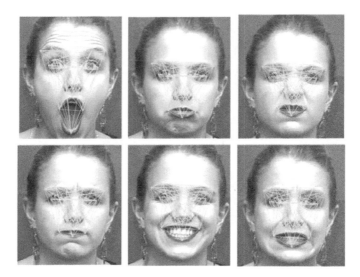

Figure 3.5: Triangle formation on the face with salient landmark points.

conditions. The features we have used for our purpose are 11 different max-min statistics of individual triangles. The features are computed by selecting three Cartesian points $(x_1, y_1), (x_2, y_2)$ and (x_3, y_3) from the combination set and the length of the three sides α, β and γ of the formed triangle is calculated using formulae 3.1, 3.2 and 3.3 respectively. We have used 11 different side length features δ_1 to δ_{11} which are computed as per equation 3.1 through 3.14 respectively. These δ_1 to δ_{11} are computed for each component set consisting of left eye, right eye, left eyebrow, right eyebrow, nose, mouth and also the face in entirety. These combined components along with a holistic feature set are used to construct a combined side-length signature representation of a face.

$$\alpha = \sqrt{(x_1 - y_1)^2 + (x_2 - y_2)^2} \qquad (3.1)$$

$$\beta = \sqrt{(x_2 - y_2)^2 + (x_3 - y_3)^2} \qquad (3.2)$$

$$\gamma = \sqrt{(x_1 - y_1)^2 + (x_3 - y_3)^2} \qquad (3.3)$$

$$\delta_1 = \frac{max(\alpha, \beta, \gamma)}{\alpha + \beta + \gamma} \qquad (3.4)$$

$$\delta_2 = \frac{min(\alpha, \beta, \gamma)}{\alpha + \beta + \gamma} \qquad (3.5)$$

$$\delta_3 = \frac{min(\alpha, \beta, \gamma)}{max(\alpha + \beta + \gamma)} \tag{3.6}$$

$$\delta_4 = \frac{max(\alpha, \beta, \gamma) - min(\alpha, \beta, \gamma)}{\alpha + \beta + \gamma} \tag{3.7}$$

$$\delta_5 = \frac{max(\alpha, \beta, \gamma) + min(\alpha, \beta, \gamma)}{\alpha + \beta + \gamma} \tag{3.8}$$

$$\delta_6 = \frac{max(\alpha, \beta, \gamma) - min(\alpha, \beta, \gamma)}{max(\alpha, \beta, \gamma) + min(\alpha, \beta, \gamma)} \tag{3.9}$$

$$\delta_7 = \frac{max(\alpha, \beta, \gamma)}{max(\alpha, \beta, \gamma) + min(\alpha, \beta, \gamma)} \tag{3.10}$$

$$\delta_8 = \frac{min(\alpha, \beta, \gamma)}{max(\alpha, \beta, \gamma) + min(\alpha, \beta, \gamma)} \tag{3.11}$$

$$\delta_9 = \frac{min(\alpha, \beta, \gamma)}{max(\alpha, \beta, \gamma) - min(\alpha, \beta, \gamma)} \tag{3.12}$$

$$\delta_{10} = \frac{min(\alpha, \beta, \gamma)}{(\alpha + \beta + \gamma) - min(\alpha, \beta, \gamma)} \tag{3.13}$$

$$\delta_{11} = \frac{max(\alpha, \beta, \gamma)}{(\alpha + \beta + \gamma) - min(\alpha, \beta, \gamma)} \tag{3.14}$$

Figure 3.6 is an example of a single image demonstrating the flow of computation of our method; this is also the image evolution counterpart of Figure 3.1.

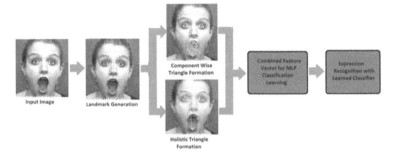

Figure 3.6: The flow of the computation for an example image.

3.5.4   Classification Learning

Learning the classifier machine is a vital step of computation for facial expression recognition. The selection of a proper classifier is extremely important to achieve a robust classifier. We have used MultiLayer Perceptron(MLP) for classification learning. The MLP we have designed has four layers: an input layer, a hidden layer, an output layer and a SoftMax layer at the end. The input layer consists of neurons equivalent to the size of an input feature vector, the hidden layer consists of 20 neurons, the output layer consists of 6 or 7 neurons depending on the number of expression classes and the SoftMax layer consists of the same number of neurons as the output layer. The SoftMax layer modifies the output layer results as a probability distribution on the number of classes of expressions. The proposed MLP is learned with scaled conjugate gradient descent backpropagation learning [23]. We have allocated 15% of the dataset for testing purposes, 15% of the dataset for validation purposes and the rest of 70% for training purposes. Results of the classification are calculated in terms of accuracy in training, testing, validation, and overall dataset. Also, we have performed a 10-fold cross-validation on each dataset. The confusion matrix showing misclassification in each class is computed for each database. False Positive (FP), False Negative (FN), True Positive (TP) and True Negative (TN) rates are calculated for each individual class of expression. The authors have also computed the False Rejection Rate (FRR), False Acceptance Rate (FAR), and Error Rate (ERR) for each database from [24].

False Acceptance Rate (FAR) is the probability of incorrect matching for a candidate image. FAR can be formulated as

$$FAR = \frac{FP}{FP + TN} \tag{3.15}$$

False Rejection Rate (FRR) signifies the total number of valid candidate images that failed to be properly identified by the system to the total number of candidate images in the database. This can be formulated as

$$FRR = \frac{FP}{FN + TP} \tag{3.16}$$

Error Rate (ERR) is the total number of misclassified images against the number of candidate images in the database. We can calculate this as

$$ERR = \frac{FP + FN}{TP + TN + FP + FN} \tag{3.17}$$

3.6 DISCUSSIONS AND PERFORMANCE COMPARISONS

We have tested the proposed method on Cohn-Kanade (CK+) [25], Japanese Female Facial Expression (JAFFE) [26], MMI [27] and Multimedia Understanding Group (MUG) [28] databases.

3.6.1 Results on CK+ Database

The extended Cohn-Kanade (CK+) dataset is an Action Unit (AU) coded database which is a sequence database comprising posed and spontaneous expressions jointly [25]. Every sequence starts from the natural expression and goes up to intense final expression. The final expression in every sequence has an associated emotion level describing the corresponding basic expression class. This database is posed by 97 different adult subjects giving seven atomic expressions: Anger(AN), Fear(FE), Contempt(CO), Disgust(DI), Happiness(HA), Sadness(SA) and Surprise(SU) thus generating 486 different sequences. We have included only the final expression of each sequence for our analysis. The proposed machine reached a staggering 99.8% accuracy in this database with 97.96% validation and 95.92% testing precessions. The confusion matrix is presented on Table 3.1 demonstrating the Fear(FE), Contempt(CO), Disgust(DI), Happiness(HA) and Sadness(SA) classes of expression, which shows an overwhelmingly 100% precessions. The true-false rate chart is given on Table 3.2, which shows that the contempt class of expression has the lowest 0.900 precessions. The FAR, FRR and ERR rates on this database are shown on Table 3.3. A 10-fold validation test is performed on CK+ database achieving 89.59% accuracy and the results on each fold is shown on Figure 3.7.

TABLE 3.1: Confusion matrix of CK+ database.[25]

	AN	CO	FE	DI	HA	SA	SU
AN	44	1	0	0	0	0	0
CO	0	18	0	0	0	0	0
DI	0	0	59	0	0	0	0
FE	0	0	0	25	0	0	0
HA	0	0	0	0	69	0	0
SA	0	0	0	0	0	28	0
SU	0	1	0	1	0	0	81

TABLE 3.2: The False Negative (FN), False Positive (FP), True Positive (TP), and True Negative (TN) Rates on CK+ database.

	FN	FP	TP	TN
AN	0.004	0	1	0.996
CO	0	0.100	0.900	1
DI	0	0	1	1
FE	0	0.038	0.961	1
HA	0	0	1	1
SA	0	0	1	1
SU	0.008	0	1	0.992

TABLE 3.3: FAR, FRR and rates on CK+ database.[24]

	FAR	FRR	ERR
AN	0	0.00352	0.00176
CO	0.09090	0	0.05000
DI	0	0	0
FE	0.03703	0	0.01923
HA	0	0	0
SA	0	0	0
SU	0	0.0080	0.00406

3.6.2 Results on JAFFE database

Japanese Female Facial Expression (JAFFE) database is a synthetic expression database incorporating the six rudimentary expressions set along with natural face images [26]. The JAFFE database comprises 213 images posed by 60 different models, and this database does not contain any sequence. Unlike the CK+, the JAFFE database is created in the Psychology Department in Kyushu University, Japan, and the apparatus for capturing the image includes setup of a camera with a semi-reflective plastic sheet. The semi-reflective sheet is placed in the front of the camera apparatus such that each model can correct her expression by looking at her reflection on the semi-reflective sheet. The expression levels are semantically rated by a different set of annotators which is also validated against a different set of analyzers thus accomplishing accurate labeling of each expression class. The performance of the proposed machine attained an impressive 98.59% accuracy in this database with 96.88% and 93.75% validation and testing precision respectively. The confusion matrix is presented in Table 3.4 demonstrating a total of three misclassifications and an overwhelmingly 100% precision accomplished in Fear(FE) and Sadness(SA) classes of expression. The true positive rate from Table 3.5 shows that the Anger, Sadness and Neutral expressions have a comparatively low precision of almost 96.60%. The authors have also computed FAR, FRR and ERR on Table 3.6. We have applied 10-fold validation testing on the JAFFE database which shows 78.12% accuracy shown in Figure 3.7.

TABLE 3.4: Confusion matrix of JAFFE database.[26]

	AN	DI	HA	FE	SA	SU	NE
AN	30	0	0	0	0	0	1
DI	0	31	0	0	0	0	0
HA	0	0	30	0	0	0	0
FE	0	0	0	30	0	0	0
SA	0	0	0	0	29	0	0
SU	1	0	0	0	1	30	0
NE	0	0	0	0	0	0	30

TABLE 3.5: The False Negative (FN), False Positive (FP), True Positive (TP), and True Negative (TN) Rates on JAFFE database.

	FN	FP	TP	TN
AN	0.005	0.032	0.967	0.994
DI	0	0	1	1
HA	0	0	1	1
FE	0	0	1	1
SA	0	0.033	0.966	1
SU	0.011	0	1	0.989
NE	0	0.0320	0.967	1

TABLE 3.6: FAR, FRR and ERR rates on JAFFE database. [24]

	FAR	FRR	ERR
AN	0.03141	0.00564	0.01887
DI	0	0	0
HA	0	0	0
FE	0	0	0
SA	0.03225	0	0.01666
SU	0	0.01081	0.00546
NE	0.03125	0	0.01612

3.6.3 Results on MMI database

The MMI database is created by Pantic et al. at Imperial College London, Twente University to meet the gap of a freely available database with Action Unit (AU) annotation in the facial expression research community [27]. The MMI database later incorporated expression level for six prototypic expressions. The MMI database is a multistage combined database containing both posed and spontaneous expressions. We have used 222 images from stage IV of this dataset for classification learning. In the MMI database, we have attained an interestingly 97.29% of overall accuracy with 96.96% and 84.84% of validation and testing accuracy respectively. The confusion matrix displayed on Table 3.7 shows a total number of 5 misclassifications and a surprisingly 100% precessions accomplished in Fear(FE), Happiness(HA) expression classes. The true positive rate from Table 3.8 indicates that the Anger(AN), Fear(FE) and Sadness(SA) expressions have comparatively low precessions rates of 95.30%, 96.00% and 96.40% respectively. The authors have also computed FAR, FRR and ERR displayed on Table 3.9. A 10-fold cross-validation shows 84.84% accuracy on MMI database and fold wise results are plotted in Figure 3.7.

TABLE 3.7: Confusion Matrix of MMI database. [27]

	AN	DI	FE	HA	SA	SU
AN	61	0	0	0	1	0
DI	0	23	0	0	0	1
FE	0	0	24	0	0	0
HA	0	0	0	45	0	0
SA	1	0	0	0	27	0
SU	2	0	1	0	0	36

TABLE 3.8: The False Negative (FN), False Positive (FP), True Positive (TP), and True Negative (TN) Rates on MMI database.

	FN	FP	TP	TN
AN	0.006	0.046	0.953	0.993
DI	0.005	0	1	0.994
FE	0	0.040	0.960	1
HA	0	0	1	1
SA	0.005	0.035	0.964	0.994
SU	0.016	0.027	0.972	0.983

TABLE 3.9: FAR, FRR and ERR rates on MMI database. [24]

	FAR	FRR	ERR
AN	0.04504	0.00659	0.02660
DI	0	0.00500	0.00251
FE	0.03846	0	0.02000
HA	0	0	0
SA	0.03465	0.00531	0.02043
SU	0.02673	0.01639	0.02162

3.6.4 Results on MUG database

The Multimedia Understanding Group (MUG) research group of Aristotle University of Thessaloniki created this expression database comprising images and some video sequences [28]. A total of 86 models had been posing for the creation of this database. The poser models are 35 female and 51 male candidates having an age range of between 20 to 35 years. This database is not fully open for the research community, but we have achieved a public copy of the database which comprises 401 frontal facial images of 26 models showing six basic expressions with some variations. Classification learning with MLP in the MUG database gave interestingly 98.25% of overall accuracy with 95% and 93.33% of validation and testing accuracy respectively. The confusion matrix is presented on Table 3.10 shows a total of 4 number misclassifications and an astonishingly 100% accuracy realized in Disgust(DI), Fear(FE) and Sadness(SA) expression classes. The true positive rate from Table 3.11 refers to a comparatively low value in Anger expression having 0.889% precession. The authors have also computed FAR, FRR and ERR displayed on Table 3.12. A 10-fold cross-validation shows 80.5% accuracy on this database with fold wise accuracy plotted in Figure 3.7.

TABLE 3.10: Confusion matrix of MUG database. [28]

	AN	DI	FE	HA	SA	SU
AN	24	0	1	0	0	0
DI	0	57	0	0	0	0
FE	0	0	71	0	0	0
HA	1	0	0	46	0	0
SA	0	0	0	0	87	0
SU	2	0	0	0	0	46

TABLE 3.11: The False Negative (FN), False Positive (FP), True Positive (TP), and True Negative (TN) Rates on MUG database.

	FN	**FP**	**TP**	**TN**
AN	0.002	0.111	0.889	0.997
DI	0	0	1	1
FE	0	0.027	0.972	1
HA	0.002	0	1	0.997
SA	0	0.011	0.988	1
SU	0.005	0.021	0.978	0.994

TABLE 3.12: FAR, FRR and ERR rates on MUG database.[24]

	FAR	**FRR**	**ERR**
AN	0.10024	0.00299	0.05689
DI	0	0	0
FE	0.02666	0	0.01369
HA	0	0.00280	0.00140
SA	0.01123	0	0.00568
SU	0.02094	0.00573	0.01346

Table 3.13 reveals that the proposed method outperformed [21] on JAFFE and CK+ database instances on one hand and near equivalent performance compared to [20] on CK+ database instances compared to the other methods. Table 3.13 shows the training, validation, testing and overall accuracy of the system in those databases.

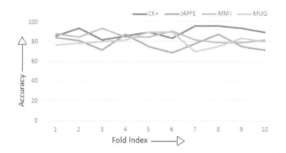

Figure 3.7: The 10-fold cross-validation results on CK+, JAFFE, MMI, and MUG database.

TABLE 3.13: Comparison of performance of MLP learning among CK+[25], JAFFE[26], MMI[27], MUG[28] databases.

	Training	Validation	Testing	Overall
CK+	100	97.95	95.91	99.08
JAFFE	100	96.87	93.75	98.59
MMI	100	96.96	84.84	97.29
MUG	100	95	93.33	98.25

TABLE 3.14: The comparision of the proposed method with different state-of-the-art facial expression recognition system.

	JAFFE Database	CK+ Database
fathallah2017[20]	95.20	99.33
Mayaa2016[21]	98.12	96.02
Proposed Method	98.59	99.08

We have also compared our method with recent standard facial expression methods shown in CK+ and JAFFE database Table 3.14.

3.7 CONCLUSIONS

The astonishingly encouraging overall accuracy of combined side-length feature-based learning shows that the system learns to affect cues morphed in the shape of the emoted face in a pose, illumination, and person independent manner. The length of three sides of the Triangle, produced as a result of joining suitably chosen landmark points, offers significant shape informative feature specification. The application of these features to achieve the task of classification into six categories reports impressive results. Future works may include exploring new shape signature descriptors relevant to triangle angle, fuzzy measure on the triangle, centers of triangles, and so forth.

3.8 ACKNOWLEDGMENTS

The authors would like to express their thankful gratitude to Prof. Maja Pantic for arranging access to the MMI database and Dr. A. Delopoulos for arranging access to the MUG database and for carrying out this work. The authors also like to thank the University Grant Commission (UGC) for providing NET-JRF fellowship

[UGC Ref-No. - 3437/(OBC)/NET-JAN-2017] for carrying out this work.

BIBLIOGRAPHY

[1] Rosalind W Picard. *Affective computing*. MIT press, 2000.

[2] Paul Ekman and Wallace V Friesen. Constants across cultures in the face and emotion. *Journal of personality and social psychology*, 17(2):124, 1971.

[3] Paul Ekman. An argument for basic emotions. *Cognition & emotion*, 6(3-4):169–200, 1992.

[4] Jeffrey F Cohn, Zara Ambadar, and Paul Ekman. Observer-based measurement of facial expression with the facial action coding system. *The handbook of emotion elicitation and assessment*, pages 203–221, 2007.

[5] Paul Viola and Michael J Jones. Robust real-time face detection. *International journal of computer vision*, 57(2):137–154, 2004.

[6] Caifeng Shan, Shaogang Gong, and Peter W McOwan. Facial expression recognition based on local binary patterns: A comprehensive study. *Image and vision Computing*, 27(6):803–816, 2009.

[7] Haitao Wang, Stan Z Li, and Yangsheng Wang. Face recognition under varying lighting conditions using self quotient image. In *Sixth IEEE International Conference on Automatic Face and Gesture Recognition, 2004. Proceedings.*, pages 819–824. IEEE, 2004.

[8] SL Happy and Aurobinda Routray. Automatic facial expression recognition using features of salient facial patches. *IEEE transactions on Affective Computing*, 6(1):1–12, 2015.

[9] Mahdi Ilbeygi and Hamed Shah-Hosseini. A novel fuzzy facial expression recognition system based on facial feature extraction from color face images. *Engineering Applications of Artificial Intelligence*, 25(1):130–146, 2012.

[10] V Gomathi, K Ramar, and A Santhiyaku Jeevakumar. A neuro fuzzy approach for facial expression recognition using lbp histograms. *International Journal of Computer Theory and Engineering*, 2(2):245, 2010.

[11] Asit Barman and Paramartha Dutta. Facial expression recognition using distance and shape signature features. *Pattern Recognition Letters*, 2017.

[12] Asit Barman and Paramartha Dutta. Texture signature based facial expression recognition using narx. In *2017 IEEE Calcutta Conference (CALCON)*, pages 6–10. IEEE, 2017.

[13] Asit Barman and Paramartha Dutta. Facial expression recognition using distance signature feature. In *Advanced Computational and Communication Paradigms*, pages 155–163. Springer, 2018.

[14] Asit Barman and Paramartha Dutta. Facial expression recognition using distance and texture signature relevant features. *Applied Soft Computing*, 77:88–105, 2019.

[15] Asit Barman and Paramartha Dutta. Influence of shape and texture features in facial expression recognition. *IET Image Processing*, 04 2019.

[16] Asit Barman and Paramartha Dutta. Influence of shape and texture features in facial expression recognition. *IET Image Processing*, 2019.

[17] Cigdem Turan, Kin-Man Lam, and Xiangjian He. Soft locality preserving map (slpm) for facial expression recognition. *arXiv preprint arXiv:1801.03754*, 2018.

[18] Anagha S Dhavalikar and RK Kulkarni. Face detection and facial expression recognition system. In *2014 International Conference on Electronics and Communication Systems (ICECS)*, pages 1–7. IEEE, 2014.

[19] Ping Liu, Shizhong Han, Zibo Meng, and Yan Tong. Facial expression recognition via a boosted deep belief network. In *Proceedings of the IEEE Conference on Computer Vision and Pattern Recognition*, pages 1805–1812, 2014.

[20] A. Fathallah, L. Abdi, and A. Douik. Facial expression recognition via deep learning. In *2017 IEEE/ACS 14th International Conference on Computer Systems and Applications (AICCSA)*, pages 745–750, Oct 2017.

[21] Veena Mayya, Radhika M. Pai, and M.M. Manohara Pai. Automatic facial expression recognition using dcnn. *Procedia Computer Science*, 93:453 – 461, 2016. Proceedings of the 6th International Conference on Advances in Computing and Communications.

[22] Georgios Tzimiropoulos and Maja Pantic. Optimization problems for fast aam fitting in-the-wild. In *Proceedings of the IEEE international conference on computer vision*, pages 593–600, 2013.

[23] Martin Fodslette Møller. A scaled conjugate gradient algorithm for fast supervised learning. *Neural networks*, 6(4):525–533, 1993.

[24] Sébastien Marcel, Mark S Nixon, and Stan Z Li. *Handbook of biometric anti-spoofing*, volume 1. Springer, 2014.

[25] Patrick Lucey, Jeffrey F Cohn, Takeo Kanade, Jason Saragih, Zara Ambadar, and Iain Matthews. The extended cohn-kanade dataset (ck+): A complete dataset for action unit and emotion-specified expression. In *2010 IEEE Computer Society Conference on Computer Vision and Pattern Recognition-Workshops*, pages 94–101. IEEE, 2010.

[26] Michael Lyons, Shigeru Akamatsu, Miyuki Kamachi, and Jiro Gyoba. Coding facial expressions with gabor wavelets. In *Proceedings Third IEEE international conference on automatic face and gesture recognition*, pages 200–205. IEEE, 1998.

[27] Michel Valstar and Maja Pantic. Induced disgust, happiness and surprise: an addition to the mmi facial expression database. In *Proc. 3rd Intern. Workshop on EMOTION (satellite of LREC): Corpora for Research on Emotion and Affect*, page 65, 2010.

[28] Niki Aifanti, Christos Papachristou, and Anastasios Delopoulos. The mug facial expression database. In *11th International Workshop on Image Analysis for Multimedia Interactive Services WIAMIS 10*, pages 1–4. IEEE, 2010.

CHAPTER 4

A Study on the Influence of Angular Signature of Landmark Induced Triangulation in Recognizing Changes in Human Emotion

Md Nasir, Paramartha Dutta, and Avishek Nandi

A Nalysis of human emotion from the sequence of face images is a vital issue for a person to recognize the changes in emotional behavior. In this work, we propose an effective method for recognition of temporal dynamic variations in human emotion from facial video frames that used a triangulation mechanism to generate triangles on the face. In our approach, angular information extracted from every triangle generated by landmark points is taken into the account as geometric features that help us to distinguish human emotion into different basic facial expressions like anger, disgust, fear, happiness, sadness, and surprise. Besides, we considered important regions on the face to get relevant geometric features that discriminate an image sequence from others. For verification of the performance of our proposed method, we experimented with our emotion recognition system on different benchmark image sequence databases like Extended Cohn-Kanade(CK+), MMI and MUG. A comparison in experimental results obtained from different databases encompasses the efficiency of the proposed method with a promising recognition rate.

4.1 INTRODUCTION

Recognition of temporal changes of human facial expressions provides a substantial impact on researchers in recent eras. Of late Affective Computing [3], which deals with intelligent recognition of human emotion, attracts the attention of the concerned research community to a large extent. It is very difficult to understand human emotions for a machine but an Affective Computation facilitates a machine to perceive, realize and amalgamate emotions. According to [4], various expressions of a human face at different levels has been regulated by varying facial activities such as the action of muscles on the face may cause changes in facial behaviors individually or in groups. Paul Ekman and Friesen [4] have placed human emotions into six different basic expression labels: anger, disgust, fear, happiness, sadness, and surprise. They introduced the Facial Action Coding System (FACS) imposing an indication of recognition of landmarks movement on the face in terms of the temporal profile of action units (AUs). Authors in [5] realized that detection of perfect landmark points on the face is a more difficult task than facial expression classification. Active Appearance Model (AAM) [2] is an assortment of both texture and shape models that provides landmark points on the face. In [6], the authors mentioned that Facial Expression Recognition System (FERS) can be developed by using two main approaches: one is a static based approach which used static images to get geometric features and another is a dynamic based approach that utilized video frames. Tracking of landmark points on the face from video frames is a more challenging task than static images due to high data dimensionality. Authors in [7],[8],[9] used static images for emotion recognition. Only finding the best geometric representation of images or sequence of images is not sufficient to classify the human emotion. To make a perfect classification of expression, a classifier needs to play an important role. Most of the approaches like [10], [11], [12] used Support Vector Machine (SVM), Artificial Neural Network (ANN) and Naïve Bayesian Classifier (NB) respectively to discriminate facial expressions into different basic labels.

Motivation: In recent decades, authors in [13], [14] and [15] applied image sequences into their recognition system to identify the temporal behavior of an expression. It is observed that the recognition system based on static images is unable to capture time-varying activities of individual expression due to a lack of intermediate frames of emotion. These thoughts motivate us to deal with image sequences to find out information about the dynamic characteristics of human emotion. Our proposed method used all frames in the transition from neutral to peak expression to bridge the information gap between static and dynamic based approaches.

Our contributions: 1) Triangle generation from landmark points: Geometric positions of landmark points on the face are identified by applying the Active Appearance Model (AAM) on image sequence. Then we have generated triangles from those relevant geometric positions which are associated with the important regions on the face: eyes, eyebrows, nose, and lips. 2) Angular signature: The triangulation mechanism is used to construct an angular signature by considering three angles of each triangle. We prepared an angular signature by taking the ratio between angles. 3) Classification on angular signature: Angular signature is formed as a prominent geometric feature with high discrimination power and is fed into Artificial Neural Network (ANN) to distinguish sequences into basic expression labels. Classification is conducted on various benchmark image sequence databases: CK+, MMI, and MUG. Angles formed by different triangles joining selected landmark points on facial images are not yet explored as an effective feature descriptor for expression identification. This is as per the best of our knowledge in the relevant domain. In our opinion, the application of such a feature descriptor is a novel initiative.

The remaining parts of this paper are organized as follows. We presented the proposed method in section 4.2. Results and discussion are described in section 4.3. In section 4.4, we provided a comparison of our results with those of other methods. In the end, the conclusion is drawn in section 4.5.

4.2 PROPOSED METHOD

Our proposed Facial Expression Recognition System (FERS) is divided into three subsystems: a) landmark identification, b) geometric feature extraction and c) emotion recognition. The workflow of our proposed method is shown in Figure 4.1. We have developed the emotion recognition system based on the dynamic approach using facial image sequences. We need to address the dynamic

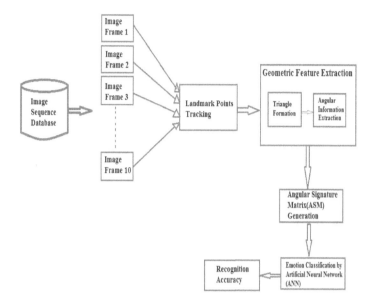

Figure 4.1: Diagram of our proposed recognition system

behavior of human emotion through our proposed system. Geometric based features are used to accumulate the information for discriminating one emotion from others. That is why the development of our recognition system is initiated with landmark identification providing geometric locations. Depending on landmark projection on facial images, we have reduced the number of frames from the sequence. We have selected the best ten frames from every sequence into consideration according to the perfect projection of landmark points on the frame where the first frame containing a neutral image and the last frame is one of the basic emotions. Figures 4.2, 4.3, 4.4, 4.5, 4.6 and 4.7 show the transition of anger, disgust, fear, happiness, sadness and surprise emotions from neutral respectively.

4.2.1 Landmark Identification

Facial muscle points on a facial image have a huge impact on the recognition of changes in the behavior of an emotion. These points are obtained by our proposed landmark identification subsystem. We have applied the Active Appearance Model [2] on image sequence to identify the geometric location of landmark points. A total sixty-eight of geometric locations are raised from every frame

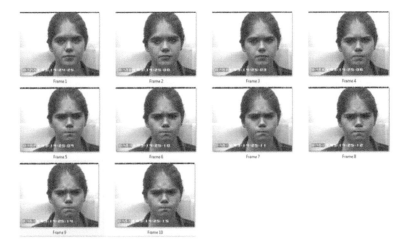

Figure 4.2: Video of 10 frames starting with neutral expression and ending with anger expression image

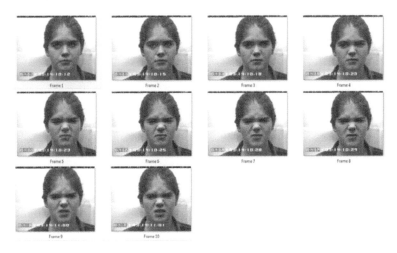

Figure 4.3: Video of 10 frames starting with neutral expression and ending with disgust expression image

in a sequence. Figure 4.8 shows the positions of those sixty-eight landmark points on single image frame. In the study of Barman and Dutta [1], it is noticed that among them only twenty-one locations are playing a crucial role to detect the changes in the behavior, and we have added an extra two points taken from the lips for further detailing. As per [1] these selected points are those which reflect maximum sensitivity for various expression types. Those crucial geometric locations are very sensitive due to the dislocation of

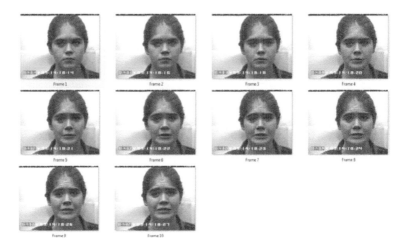

Figure 4.4: Video of 10 frames starting with neutral expression and ending with fear expression image

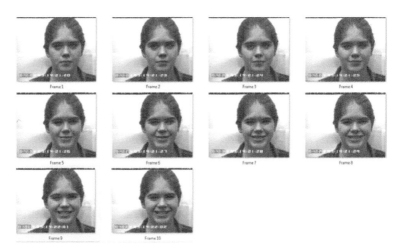

Figure 4.5: Video of 10 frames starting with neutral expression and ending with happiness expression image

major portions of the face and it is measured through analyzing the movement of landmark points over the frames in the sequence. Twenty-three points are considered from four major components of face: eight points are taken from eyes, six points are from eyebrows, and three and six points are selected from nose and lips respectively. Figure 4.9 shows crucial landmark points on the face extracted from a single frame.

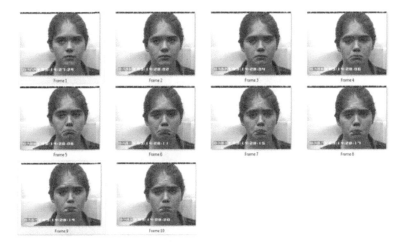

Figure 4.6: Video of 10 frames starting with neutral expression and ending with sadness expression image

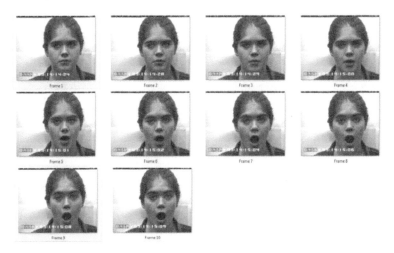

Figure 4.7: Video of 10 frames starting with neutral expression and ending with surprise expression image

4.2.2 Geometric Feature Extraction

The most powerful component of the Facial Expression Recognition System (FACS) is feature extraction that helps us to represent a sequence to be classified with minimum error. Our proposed approach used a triangulation mechanism to extract geometric features from every frame of the image sequence. Total $n = \binom{23}{3} = 1771$ number of all possible triangle shapes are

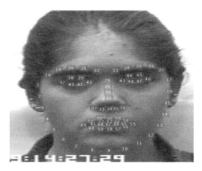

Figure 4.8: Geometric locations of sixty-eight landmark ponits

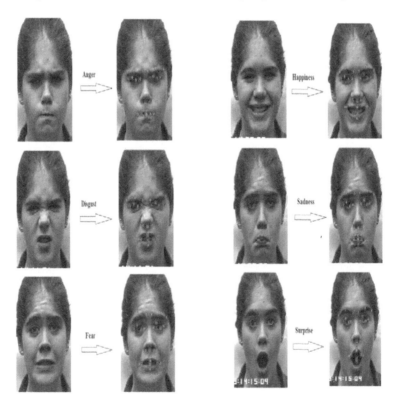

Figure 4.9: Crucial landmark detection from image frame

generated for each frame by joining every combination of three landmark points out of twenty-three points associated with major components of the face. For every triangle, we computed three angles to measure angular signature. In this tactic, we generated angular vector of size $m \times n = 10 \times 1771 = 17710$ that indicates

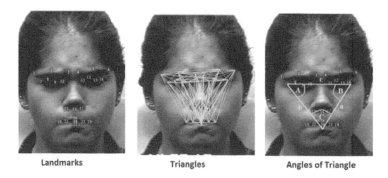

Figure 4.10: Angular information generation from landmarks for anger emotion

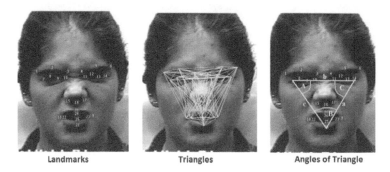

Figure 4.11: Angular information generation from landmarks for disgust emotion

a single sequence with containing angle ratio. Here $m = 10$ is a number of frames used in every sequence. Figures 4.10, 4.11, 4.12, 4.13, 4.14 and 4.15 show angles generation from an image frame for all different types of emotion.

4.2.2.1 Formation of Angular Signature Matrix (ASM) by Triangulation mechanism

Given every three geometric locations $(x_1, y_1), (x_2, y_2)$ and (x_3, y_3) of landmark points shown in Figure 4.16 depicted from Figure 4.15 as for showing the computation of all equations 4.1, 4.2, 4.3, 4.4, 4.5 and 4.6, we have calculated lengths of all sides of the triangle: a, b and c by using Euclidean distance formulations given below

$$a = \sqrt{(y_3 - y_2)^2 + (x_3 - x_2)^2} \qquad (4.1)$$

$$b = \sqrt{(y_3 - y_1)^2 + (x_3 - x_1)^2} \qquad (4.2)$$

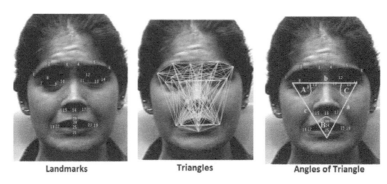

Landmarks   Triangles   Angles of Triangle

Figure 4.12: Angular information generation from landmarks for fear emotion

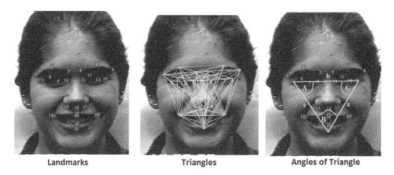

Landmarks   Triangles   Angles of Triangle

Figure 4.13: Angular information generation from landmarks for happiness emotion

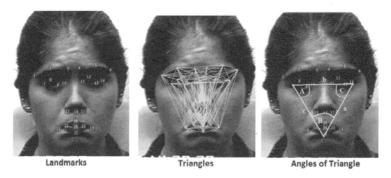

Landmarks   Triangles   Angles of Triangle

Figure 4.14: Angular information generation from landmarks for sadness emotion

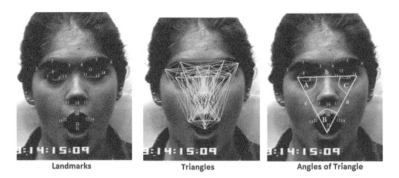

Figure 4.15: Angular information generation from landmarks for surprise emotion

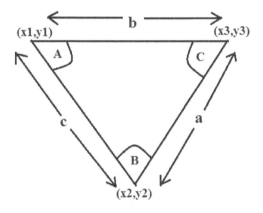

Figure 4.16: Triangle depicted from Figure 4.15

$$c = \sqrt{(y_2 - y_1)^2 + (x_2 - x_1)^2} \qquad (4.3)$$

After computing 3 sides of the triangle, all angles of A, B and C are calculated by following formulas

$$A = \cos^{-1}\left(\frac{b^2 + c^2 - a^2}{2 \times b \times c}\right) \qquad (4.4)$$

$$B = \cos^{-1}\left(\frac{a^2 + c^2 - b^2}{2 \times a \times c}\right) \qquad (4.5)$$

$$C = \cos^{-1}\left(\frac{a^2 + b^2 - c^2}{2 \times a \times b}\right) \qquad (4.6)$$

As we have used angular information as a geometric feature it stored in feature vector $\vee$ representing a single frame in a sequence and it is measured by the given equation

$$\vee_i = \frac{max(A,B,C) - min(A,B,C)}{A+B+C} \tag{4.7}$$

Finally, we have formed Angular Signature Matrix (ASM) of size $m \times n$ that denotes an image sequence and it is calculated by given below

$$ASM_k = \sum_{m=1}^{10} \sum_{n=1}^{1771} \vee_n^m \tag{4.8}$$

Here j and i are the indices of sequence number and triangle number respectively.

4.2.3 Emotion Classification

In our system, changes in human emotion concerning time are recognized by taking the image sequences as the input of the Artificial Neural Network (ANN) classifier. Figure 4.17 shows the architecture of the ANN for the sake of ready understanding. Each image in a sequence is represented by a vector of length 1771 constructed from the angular signature. As we considered 10 images for every sequence, we got a single vector of length 17710 after concatenating all 10 vectors representing a sequence fed into the classifier to classify emotions. The network consists of 3 layers: one input layer containing 17710 neurons, one hidden layer having 10 hidden neurons and one output layer containing 6 output neurons. Input of the network is image sequence represented by the vector of length 17710 and the network yields six basic emotions: anger, disgust, fear, happiness, sadness, and surprise as output. Here we have used a scaled conjugate gradient training algorithm for the network that adjusts weight and bias values till finding the minimum error calculated by Mean Square Error (MSE). We have got the classification results after applying the training algorithm exactly 50 times.

4.3 RESULTS AND DISCUSSION

Performance of our proposed method is examined by using the benchmark image sequence databases: CK+ [16], MMI [17] and MUG [18]. We have utilized all databases for emotion recognition by dividing those into three parts: 70% of total sequences are consumed for training purposes, 30% sequences are expended for validation and the remaining 30% are reserved for testing purposes. We have deliberated the angular signature as a feature

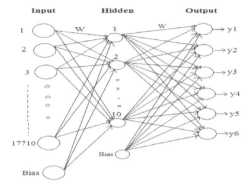

Figure 4.17: Architecture of ANN

vector extracted from ten frames of every sequence. Resultant feature vectors for every sequence are fed into MultiLayer Perceptron (MLP) to differentiate one sequence from others. As a classification result, we provide a confusion matrix along Recall, Precision and F-score measurement of every individual emotion label for all sequence databases: CK+, MUG, and MMI. Recall is calculated by following formula, $Recall = \frac{TP}{TP+FN}$, Precision is computed by $Precision = \frac{TP}{TP+FP}$ and F-score is calculated by $F - score = \frac{2 \times Precision \times Recall}{Precision + Recall}$. Here TP(True Positive) = number of image sequences are correctly identified by the classifier, FN(False Negative) = number of image sequences is incorrectly identified with other class and FP(False Positive) = number of image sequences are incorrectly identified as an actual class in the database.

4.3.1 Experiment on CK+ Database

The extended Cohn-Kanade(CK+) image sequence database contains both posed and non-posed expressions of 210 adults taken by a Panasonic AG-7500 video recorder with a Horita synchronized time-code generator. All adults were age 18-50 years, 69% of total adults were women, 31% men, 81% Euro-American, 13% African-American and 6% other groups. Image sequences were 8-bit grayscale and sequence duration were 9-60 frames/sequence [16]. In our experiment, we have used only posed expressions to analyze the characteristics of individual emotions. A total of 327 image sequences are taken into the account with seven emotions: anger (AN), contempt(CON), disgust(DI), fear(FE), happiness(HA), sadness(SA) and surprise(SU) to justify the performance

---

Algorithm 1: Emotion Classification Algorithm using Angular Signature

---

Input: Sequences of images S with corresponding emotion labels.

Output: Emotions are identified with their basic labels.

1 begin
2 for each sequence s ∈ S
3 for each frame f ∈ s
4 AAM fitting Model is applied on f to extract landmark points;
5 Compute all possible triangles T;
6 for each triangle t ∈ T
7 Find out 3 angles of t;
8 Compute Angular Signature and store it into vector v_f;
9 end
10 end
11 Concatenate all computed vector v_f into single vector V_s ;
12 end
13 Finally this feature vector V_s is fed into ANN;
14 end

---

of our emotion recognition system. Sequences of six different types of emotion of a person taken from CK+ database are shown in Figures 4.18,4.19,4.20,4.21,4.22 and 4.23. Table 4.1 shows frequency of individual emotion. The confusion matrix shows the discrimination power of the angular signature through the recognition rate, and it is shown in Table 4.2. Anger, disgust, sadness, and surprise – these emotions are perfectly recognized with a 100% accuracy rate. Contempt is identified in 16 image sequences properly, but 2 image sequences are confused with happiness and sadness. Fear is recognized in 24 image sequences, and 1 is misclassified with surprise. The overall performance of our proposed system on the CK+ database measured in terms of the average recognition rate is 98.77%. Table 4.3 shows the Recall, Precision, and F-score of our classification results.

4.3.2   Experiment on MUG Database

Our method performed the recognition task on images of the MUG database collected from 86 several subjects. Among them, 35 female instances and 51 male instances participated in this dataset. They ranged in age between 20 to 35 years. The sequence of images was captured by the camera at a rate of 19 frames/second.

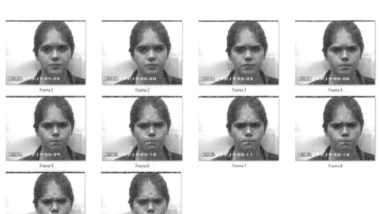

Figure 4.18: Image sequence of anger emotion in CK+ Database

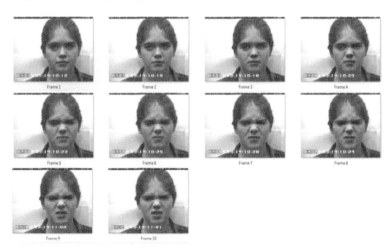

Figure 4.19: Image sequence of disgust emotion in CK+ Database

Each image in the sequence was stored as a jpg format having 896×896 pixels resolution. The subjects are not wearing glasses and there are no occlusions except for the presence of few hair instances on the face. We have used a total of 801 image sequences (561 sequences for training, 120 sequences for validation purpose and the remaining 120 sequences for testing) to classify the six basic facial expressions: anger(AN), disgust(DI), fear(FE), happiness(HA), sadness(SA) and surprise(SU). The frequency of individual emotions in the MUG database is shown in Table 4.4. As examples Figures 4.24,4.25,4.26,4.27,4.28 and 4.29 show sequences

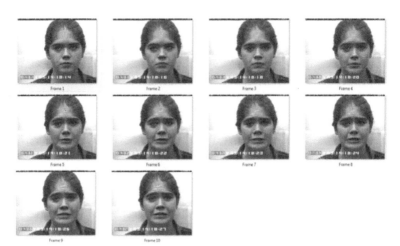

Figure 4.20: Image sequence of fear emotion in CK+ Database

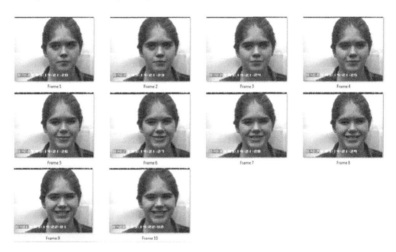

Figure 4.21: Image sequence of happy emotion in CK+ Database

of all six types of basic emotions taken from a particular subject of the MUG database. Another justification of our classification results on the MUG database is shown in Table 4.6.

Table 4.5 shows the recognition performance of our proposed angular signature on the MUG database. Here the fear expression is classified precisely with a 100% rate of accuracy. The anger emotion is recognized with a 99.32% classification rate. Disgust, sadness, and surprise are acknowledged with recognition rates of 98.29%, 98.49% and 99.31% respectively. The lowest recognition rate of 96.26% is found in the happiness expression, and the average

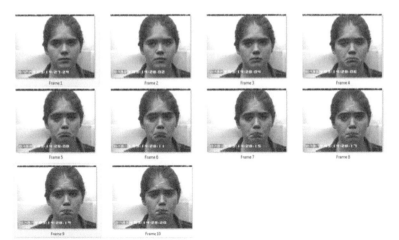

Figure 4.22: Image sequence of sadness emotion in CK+ Database

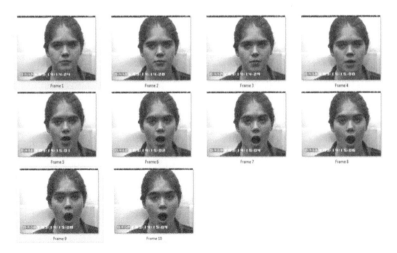

Figure 4.23: Image sequence of surprise emotion in CK+ Database

accuracy rate of 98.75% is observed on the MUG database, which is quite impressive.

4.3.3 Experiment on MMI Database

We have also experimented with our method on the MMI image sequence database, which is a very challenging task to recognize the changes in emotion by dealing with the images of this database. In this video database, it is found that both frontal image and side view images are present in every single frame. Image data was

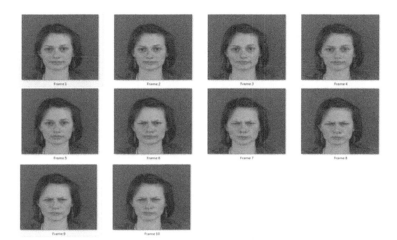

Figure 4.24: Image sequence of anger emotion in MUG Database

organized in session units of audio-visual recording. Video clips were taken from 28 different subjects, and all expressions were captured in two attempts; at first subjects were wearing glasses and subsequently without wearing glasses. We prepared the sequences by taking only frontal images and removed those sequences from this database which has no emotion label as well as whose landmarks that are not well fitted on faces. After preprocessing the images, we have found 202 sequences among a total of 236 videos available in the MMI database. The individual frequency of emotion on MMI sequence data is shown in Table 4.7. Sequences of six different types of emotions of a person taken from MMI database are shown in Figures 4.30,4.31,4.32,4.33,4.34 and 4.35. The discrimination power of our proposed method on the classification of emotion is justified by Table 4.9.

TABLE 4.1: Frequency of individual emotions on CK+ Database

Emotion	Number of sequences
Anger(AN)	45
Contempt(CON)	18
Disgust(DI)	59
Fear(FE)	25
Happiness(HA)	69
Sadness(SA)	28
Surprise(SU)	83

TABLE 4.2: Confusion Matrix on CK+ Database

	AN	CON	DI	FE	HA	SA	SU
AN	45	0	0	0	0	0	0
CON	0	16	0	0	1	1	0
DI	0	0	59	0	0	0	0
FE	0	0	0	24	0	0	1
HA	0	0	0	1	68	0	0
SA	0	0	0	0	0	28	0
SU	0	0	0	0	0	0	83

TABLE 4.3: The Recall, Precision and F-score of angular signature on CK+ Database using ANN

	Recall	Precision	F-score
An	1	1	1
CON	0.89	1	0.94
DI	1	1	1
FE	0.96	0.96	0.96
HA	0.99	0.99	0.99
SA	1	0.97	0.98
SU	1	0.99	0.99

TABLE 4.4: Frequency of individual emotions on MUG Database

Emotion	Number of Sequences
Anger(An)	149
Disgust(DI)	117
Fear(FE)	150
Happiness(HA)	107
Sadness(SA)	133
Surprise(SU)	145

TABLE 4.5: Confusion Matrix on MUG Database

	AN	DI	FE	HA	SA	SU
AN	148	0	1	0	0	0
DI	0	115	0	0	0	2
FE	0	0	150	0	0	0
HA	3	1	0	103	0	0
SA	1	0	0	0	131	1
SU	0	0	0	1	0	144

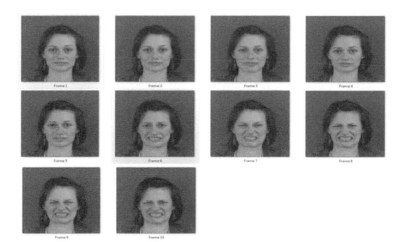

Figure 4.25: Image sequence of disgust emotion in MUG Database

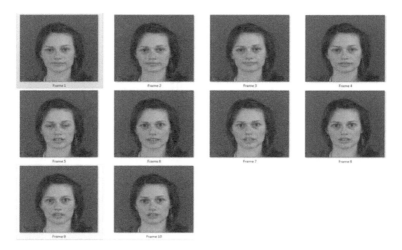

Figure 4.26: Image sequence of fear emotion in Database

Those 202 sequences are distributed according to the ratio 7: 1.5: 1.5 for training, validation and testing purposes respectively. Performances on the MMI image sequences are shown in Table 4.8 through the confusion matrix. Our method achieved an average accuracy rate of 95.54% with the lowest recognition rate of 92.85% in the case of fear expression and highest recognition rate of 97.61% in the case of happiness. It is noticed that the performance of our proposed angular signature on MMI is lower than the performances on the other two databases of CK+ and MUG. Figure 4.36 shows

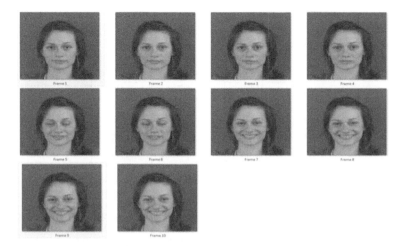

Figure 4.27: Image sequence of happiness emotion in MUG Database

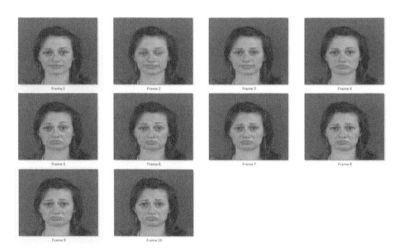

Figure 4.28: Image sequence of sadness emotion in MUG Database

the comparison in individual recognition rates of expressions under three different benchmark databases: CK+, MUG, and MMI.

4.4 COMPARISON WITH OTHER WORK

We compared our results with other existing works that experimented exclusively on the CK+database. In Table 4.10, it is noticed that the first four methods used a sequence-based approach to show recognition results reported in [19]. Our proposed method

TABLE 4.6: The Recall, Precision and F-score of angular signature on MUG Database using ANN

	Recall	Precision	F-score
AN	0.99	0.97	0.98
DI	0.98	0.99	0.98
FE	1	0.99	0.99
HA	0.96	0.99	0.97
SA	0.98	1	0.99
SU	0.99	0.98	0.98

TABLE 4.7: Frequency of individual emotions on MMI Database

Emotion	Number of Sequences
Anger(An)	31
Disgust(DI)	32
Fear(FE)	28
Happiness(HA)	42
Sadness(SA)	28
Surprise(SU)	41

TABLE 4.8: Confusion Matrix on MMI Database

	AN	DI	FE	HA	SA	SU
AN	30	1	0	0	0	0
DI	1	30	0	1	0	0
FE	1	0	26	0	0	1
HA	1	0	0	41	0	0
SA	1	0	0	0	27	0
SU	2	0	0	0	0	39

TABLE 4.9: The Recall, Precision and F-score of angular signature on MMI Database using ANN

	Recall	Precision	F-score
AN	0.97	0.83	0.89
DI	0.94	0.97	0.95
FE	0.93	1	0.96
HA	0.98	0.98	0.98
SA	0.96	0.96	0.96
SU	0.95	0.95	0.95

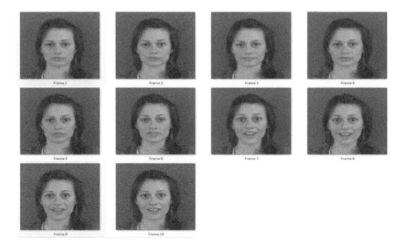

Figure 4.29: Image sequence of surprise emotion in MUG Database

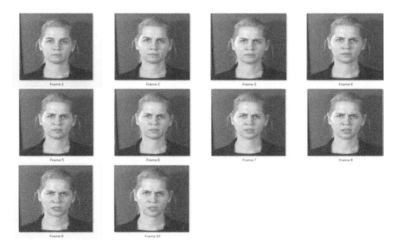

Figure 4.30: Image sequence of anger emotion in MMI Database

also used the sequence-based approach and obtained results were compared with those four [19], [20], [21], [22] methods to check the discrimination power of every method. Table 4.10 shows that our proposed approach achieved the highest average recognition results in 98.77% compared to other existing approaches.

4.5 CONCLUSION

We incorporated in this paper an automatic facial expression recognizer based on a dynamic approach to distinguish human emotion

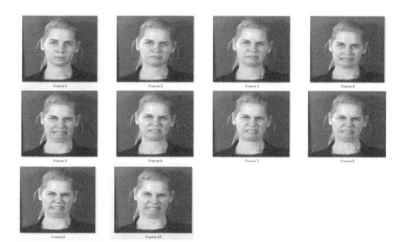

Figure 4.31: Image sequence of disgust emotion in MMI Database

Figure 4.32: Image sequence of fear emotion in MMI Database

through exploring time-varying behaviors of facial expression. Our proposed feature extraction technique is tested on various image sequence databases by presenting a triangulation mechanism. In our experiment, it is observed that angular information extracted by our method as a feature set shows the capabilities of discovering both maximal similarity within the same emotion and minimal similarity within different emotions. Further, we comprise all the recognition results of several benchmark databases; CK+, MUG, and MMI are shown in Figure 4.37. The efficiency of our system is justified by showing the average recognition rates of 98.77%,

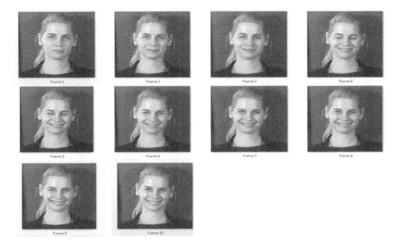

Figure 4.33: Image sequence of happiness emotion in MMI Database

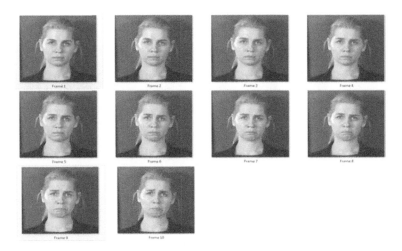

Figure 4.34: Image sequence of sadness emotion in MMI Database

98.75% and 95.54% on CK+, MUG, and MMI respectively. Comparison of our results on the CK+ database with other existing works encourages us to do the comparison for other databases: MMI and MUG and with other state-of-the-art methods for future work.

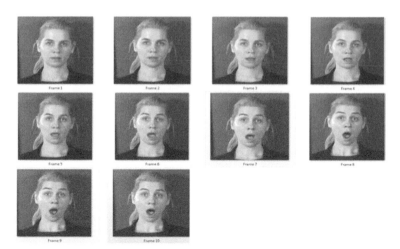

Figure 4.35: Image sequence of surprise emotion in MMI Database

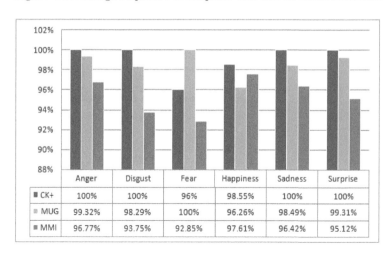

	Anger	Disgust	Fear	Happiness	Sadness	Surprise
■ CK+	100%	100%	96%	98.55%	100%	100%
▨ MUG	99.32%	98.29%	100%	96.26%	98.49%	99.31%
■ MMI	96.77%	93.75%	92.85%	97.61%	96.42%	95.12%

Figure 4.36: Comparison in recognition rate of individual expression under CK+,MUG and MMI sequence databases

4.6   ACKNOWLEDGMENT

The authors are very thankful to Dr. A. Delopoulos for providing the MUG database and Prof. Maja Pantic for providing the MMI database free for doing this work. The authors are also thankful to the Department of Computer and System Sciences, Visva-Bharati, Santiniketan for the infrastructure support. The authors gratefully

TABLE 4.10: Comparison of our results with other earlier work on CK+ Database

Input	Feature Type	Method	No of sequences	No of classes	Recog-nizer	Avg Recog-nition Rate
Image Sequence	Geometric based	Wan et al.[20]	480	6	SVM	80.00%
		Saeed et al. [21]	327	7	SVM	83.00%
		Mohammadian et al. [22]	321	7	SVM +HMM	83.34%
		Yaddaden et al. [19]	309	6	KNN	88.34%
			309	6	SVM	92.54%
Image Sequence	Geometric based	**Our approach**	**327**	**7**	**MLP**	**98.77%**

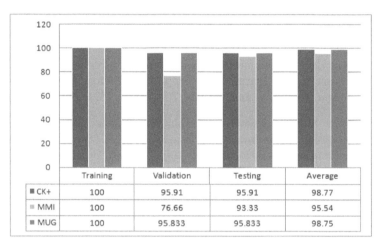

	Training	Validation	Testing	Average
■ CK+	100	95.91	95.91	98.77
▨ MMI	100	76.66	93.33	95.54
■ MUG	100	95.833	95.833	98.75

Figure 4.37: comparison of different recognition rates on CK+, MMI and MUG sequence databases

acknowledge the financial support of DST-INSPIRE Fellowship (INSPIRE Reg. no. IF160285, Ref. No.: DST/INSPIRE Fellow-ship/[IF160285]) for pursuing Doctoral Research in Department of Science and Technology, Ministry of Science and Technology, Government of India.

BIBLIOGRAPHY

[1] Asit Barman and Paramartha Dutta. Facial expression recog-nition using distance and shape signature features. *Pattern Recognition Letters*, 2017.

[2] Georgios Tzimiropoulos and Maja Pantic. Optimization problems for fast aam fitting in-the-wild. In *Proceedings of the IEEE international conference on computer vision*, pages 593–600, 2013.

[3] Rosalind W Picard. Affective computing: challenges. *International Journal of Human-Computer Studies*, 59(1-2):55–64, 2003.

[4] E Friesen and Paul Ekman. Facial action coding system: a technique for the measurement of facial movement. *Palo Alto*, 3, 1978.

[5] Jacob Whitehill, Marian Stewart Bartlett, and Javier R Movellan. Automatic facial expression recognition. *Social emotions in nature and artifact*, 88, 2013.

[6] Deepak Ghimire, Sunghwan Jeong, Joonwhoan Lee, and San Hyun Park. Facial expression recognition based on local region specific features and support vector machines. *Multimedia Tools and Applications*, 76(6):7803–7821, 2017.

[7] Bilge Süheyla Akkoca and Muhittin Gökmen. Facial expression recognition from static images. In *2014 22nd Signal Processing and Communications Applications Conference (SIU)*, pages 1291–1294. IEEE, 2014.

[8] Naveen Kumar HN, S Jagadeesha, and Amith K Jain. Human facial expression recognition from static images using shape and appearance feature. In *2016 2nd International Conference on Applied and Theoretical Computing and Communication Technology (iCATccT)*, pages 598–603. IEEE, 2016.

[9] Zhiding Yu and Cha Zhang. Image based static facial expression recognition with multiple deep network learning. In *Proceedings of the 2015 ACM on International Conference on Multimodal Interaction*, pages 435–442. ACM, 2015.

[10] Yuanyuan Ding, Qin Zhao, Baoqing Li, and Xiaobing Yuan. Facial expression recognition from image sequence based on lbp and taylor expansion. *IEEE Access*, 5:19409–19419, 2017.

[11] Seyed Mehdi Lajevardi and Margaret Lech. Facial expression recognition from image sequences using optimized feature selection. In *2008 23rd International Conference Image and Vision Computing New Zealand*, pages 1–6. IEEE, 2008.

[12] Zheng-xing SUN and Wen-hui XU. Facial expression recognition based on local svm classifiers. *CAAI Transactions on Intelligent Systems*, 5, 2008.

[13] Deepak Ghimire and Joonwhoan Lee. Geometric feature-based facial expression recognition in image sequences using multi-class adaboost and support vector machines. *Sensors*, 13(6):7714–7734, 2013.

[14] P Tamil Selvi, P Vyshnavi, R Jagadish, Shravan Srikumar, and S Veni. Emotion recognition from videos using facial expressions. In *Artificial Intelligence and Evolutionary Computations in Engineering Systems*, pages 565–576. Springer, 2017.

[15] Garima Sharma and Shilpi Gupta. Emotion detection in sequence of images using advanced pca with svm. In *2014 5th International Conference-Confluence The Next Generation Information Technology Summit (Confluence)*, pages 686–690. IEEE, 2014.

[16] Patrick Lucey, Jeffrey F Cohn, Takeo Kanade, Jason Saragih, Zara Ambadar, and Iain Matthews. The extended cohn-kanade dataset (ck+): A complete dataset for action unit and emotion-specified expression. In *2010 IEEE Computer Society Conference on Computer Vision and Pattern Recognition-Workshops*, pages 94–101. IEEE, 2010.

[17] Michel Valstar and Maja Pantic. Induced disgust, happiness and surprise: an addition to the mmi facial expression database. In *Proc. 3rd Intern. Workshop on EMOTION (satellite of LREC): Corpora for Research on Emotion and Affect*, page 65. Paris, France, 2010.

[18] Niki Aifanti, Christos Papachristou, and Anastasios Delopoulos. The mug facial expression database. In *11th International Workshop on Image Analysis for Multimedia Interactive Services WIAMIS 10*, pages 1–4. IEEE, 2010.

[19] Yacine Yaddaden, Mehdi Adda, Abdenour Bouzouane, Sebastien Gaboury, and Bruno Bouchard. Facial expression recognition from video using geometric features. 2017.

[20] Chuan Wan, Yantao Tian, and Shuaishi Liu. Facial expression recognition in video sequences. In *Proceedings of the 10th World Congress on Intelligent Control and Automation*, pages 4766–4770. IEEE, 2012.

[21] Anwar Saeed, Ayoub Al-Hamadi, Robert Niese, and Moftah Elzobi. Frame-based facial expression recognition using geometrical features. *Advances in Human-Computer Interaction*, 2014:4, 2014.

[22] Amin Mohammadian, Hassan Aghaeinia, and Farzad Towhidkhah. Video-based facial expression recognition by removing the style variations. *IET image processing*, 9(7):596–603, 2015.

CHAPTER 5

A Behavioural Model for Persons with Autism Based on Relevant Case Study

Rudranath Banerjee, Sourav De, and Shouvik Dey

THe research work aims to predict the behavioural patterns of children with Autism Spectrum Disorder (ASD) and create a standard behavioural model through a case study based on the Indian Scale for Assessment of Autism (ISAA) tool. A dynamically updated data processing system on a digital platform named National Autistic Information System-India (NAISI) has been designed to observe and evaluate the overall behavioural and performance patterns of persons with ASD on a real-time basis.

An observational and prospective study of NAISI in children with ASD was conducted at four centres (Secunderabad, Kolkata, Noida and Lajpatnagar) of National Institute for the Empowerment of Persons with Intellectual Disabilities (NIEPID), India. The NAISI system consisted of technical inputs, clinical inputs, classroom observation, clinical and technical inputs, personal intervention with the students, information of the students from official records, their respective class teachers and caregivers.

A total number of 60 children with ASD participated in the NAISI study. Their blood pressure and heart rates were recorded using clinical (clinical devices) technical (sensors) inputs. The results of the two inputs were then compared and found similar. After undergoing several other processes, the hypothesis of behavioural pattern was also successfully proven by comparing the emotional changes with respect to all the parameters on a real-time basis. An alternative behavioural model was designed that was inclusive of a child's emotions, likes and dislikes and surrounding environment.

This research work through the behavioural model helps to recognise the exact emotions of the children, have knowledge about their likes and dislikes and provide positive and functional environments for them to thrive.

Keywords: Autism, Behavioural model, Case study, NAISI, Sensors, FER

5.1   INTRODUCTION

Autism Spectrum Disorder (ASD) is a permanent neurological growth disorder that develops in infancy resulting mostly in social communication impairment and behavioural deficiency [1]. The disorder can range from mild to severe depending on where the person belongs in the spectrum. Recent studies suggest that approximately 1 out of every 59 individuals suffers from ASD [2]. The frequency and permanent nature of the disorder entails lifelong care as well as medical, educational and social support. While there have been certain advancements in the first two domains, there is still a lot of progress to be made in terms of social acceptance of the Autistic people. Their social uneasiness, unpredictable behaviour, problems in communication and trust issues are some of the reasons why the gap in social exchange between them and the others still exist.

Recent Autism researches reveal that behavioural prediction and analysis in a naturalistic setting can be an important and cost effective solution to the problem. As it is, Autistic behavioural patterns are different from the behavioural patterns of other people. They are unpredictable and subject to frequent changes. Thus interaction with such people becomes a challenge and even more so because unlike others, they cannot express their thoughts and emotions clearly. The behaviour of any person depends on one's emotions, surrounding environment, likes and dislikes and so on. Thus, behavioural analysis in Autism research is a complicated study that requires a detailed evaluation of the above mentioned parameters and their respective equation with an individual's behaviour.

Emotion recognition (ER) as a part of behavioural studies is a focus of recent Autism research. Emotions can be classified into three states: basic or primary, complex or secondary and compound or mixed emotions. Autistic individuals incorporate all three states in different situations. Those emotions can range from happiness to sad to anger to more subtle emotions. These subtle or mixed emotions are often difficult to predict because persons with Autism do not always express them either through body language or verbal means. Thus emotion recognition models are important in Autism research. Psychologists often use three concepts to recognise one emotion from the other:

- Valence which is appositive or a negative emotion like happiness or fear

- Arousal which is a calming or an exciting emotion. For example anger is considered high arousal while sad is low arousal

- Dominance which ranges from control to submission. For example admiration is high dominance while fear is low dominance

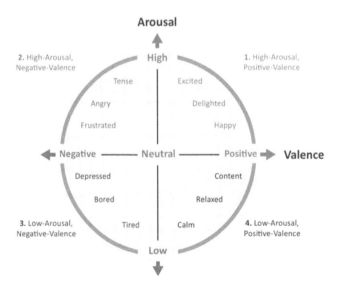

Figure 5.1: Recognizing Emotion through Valence, Arousal and Dominance

In the course of this study however, it has been found that the standard emotional models like Figure 5.1 are based only on basic emotions and do not include complex or compound emotions. This makes the models incomplete as Autistic emotions incorporate all three emotional states. Understanding Autistic behaviour entails not only the recognition of emotions but also knowing a person's likes and dislikes and the suitable environment in which an Autistic person can be comfortable. Thus standard behavioural models should include all these parameters along with emotion recognition. Unfortunately, no such model has been found during the research of the current work.

This research work based on a case study therefore proposes an alternative and unique behavioural model containing different influential parameters of behaviour such as emotions, likes and dislikes of an Autistic person and the surrounding environment. The model has been designed by using both psychological and technical expertise. The psychological discourse has been provided by four centres of NIEPID (National Institute for the Empowerment of Persons with Intellectual Disabilities, formerly National Institute for the Mentally Handicapped) at Secunderabad, Noida, Lajpatnagar and Kolkata from extensive case studies on more than 60 Autistic children between ages 3 to 20. In order to develop the behavioural model, this work takes into consideration certain hypotheses and the case study helps to prove them. They are:

1a. The behaviour of an Autistic child is directly proportional to the surrounding environment.

1b. Although each person's likes and dislikes are different from the other, there is a pattern to the overall likes and dislikes of the children involved in the case study.

2. The range of a person's heart rate varies according to one's different behavioural changes.

3. The general behaviour of an Autistic child may vary from one's classroom behaviour.

4. The behavioural changes of a child are often dependent on performing activities.

The whole work has been divided into four sections. The first section is a review of related works. Certain relevant works on behavioural analysis have been reviewed here. The next section is the methodology of the case study which details each phase of the case study. The third section results and discussion elaborate on the results of the case study and how they prove the hypotheses. The concluding section is an overview of the whole work.

5.2 REVIEW OF RELATED WORKS

Research on this study has found that works related to behavioural analysis are specific in nature. A vast majority of works deal with emotion recognition through facial expressions. Facial expressions are particularly effective as non-verbal means of communication. However their function to predict an Autistic person's behaviour is limited. Difficulty in facial emotional expression is a key to Autistic individuals[3]. This is supported by works using electromyography (EMG) or behavioural analysis of images or videos which suggest that the emotional expressions of Autistic people are less natural in both naturalistic [[4],[5]] as well as experimental [[6],[7]] situations. Hence, they are of lesser communicative value[8].

In addition to the evaluation of spontaneous facial expressions, researches have also worked on voluntary use of facial expressions of Autistic individuals [[9], [10],[11]]. Studies reveal that for typically developed (TD) people, even voluntary expressions are often effective tools of sociability[12]. However, persons with ASD do not retain the ability to produce voluntary expressions on their own. They can only do so with the intervention of an external model which can then be referred to as imitation or mimicry of expressions. This creates disparity between their real emotions and their imitated emotions [[13],[14],[15]] causing it to be even more difficult to predict behavioural patterns. Studies also reveal younger children with ASD are not capable of imitating facial expressions as well as the adults [16]. This ability only grows over time with age and maturity. However, posed emotions were found to be impaired across all age groups. Hence, facial expression recognition (whether spontaneous, imitated or posed) are not very fruitful in overall behavioural pattern recognition.

Apart from Facial Expression Recognition, researches have also focussed on use of wearable biosensors for behavioural pattern analysis. One such device is an actigraph unit or an actimetry sensor that can be used on the wrist like a watch and used to measure motor activities of an Autistic person. The data recorded is used to identify periods between idleness to energetic activities[17]. Studies on actigraphy report that Autistic individuals spend more time in idleness than in enthusiastic activities [18]. However, the relation between this information and behaviour analysis is confusing. Idleness and enthusiastic activities both point to states of happiness and contentment. It is indicative of a fixed emotional state and its corresponding behavioural exertion, excluding all other states. Actigraph units are also used to detect repetitive motor movements [[19], [20],[21]]. Although this is a better means for behavioural analysis (repetitive motor movements often highlight states of agitation or hyper-activeness), they do not include the other necessary parameters of behavioural pattern recognition like the surrounding environment and personal likes and dislikes.

There are certain other deep observational tools such as caregiver measurements of behavioural change in ASD. Two of them are Aberrant Behaviour Checklist–Community (ABC) [[22],[23]] used to measure general behaviours, and the Social Responsiveness Scale 2 (SRS-2) used to measure social behaviour [24]. The retrospective nature of the measurements may reduce accuracy especially in the absence of interventions. Therefore, it may be difficult to detect the current behavioural patterns. The Janssen Autism Knowledge Engine (JAKE) system which uses standard physiological and psychological tools to identify symptoms of ASD is an effective study system to report behaviours but does not include individual preferences or the effect of environment on the subjects.

Current researches in Autism often involve strategies based on Applied Behaviour Analysis (ABA) which identifies environmental parameters for social behaviour. Telehealth is one such strategic ABA based communication system used to aid in education and treatment of health. Research findings on Telehealth point out that it is a cost-effective ABA system for individuals with ASD especially in rural areas [25].

Studies utilizing Telehealth have worked on the functional analysis of ASD participants through video conference [26] and didactic teaching and training of students through Skype [27]. There are also studies on home-based training of ASD participants on functional communication by their parents through videos [28] and online training websites such as IMPACT [29]. Similarly, research on positive behavioural support or behaviour management has utilized Telehealth videoconferences with parents as interventionists and Autistic individuals as participants. These training sessions are mostly based on parental knowledge of ASD as well as evidence-based practices [30].

Telehealth practices are also used for written instructions through e-mail by interventionists [31], for experiments on preference assignments on paired choices through video conferences [32], for naturalistic teaching with parental implementation of communication strategies [33] and without any interventions [34]. Sometimes there are also self-directed lessons through videoconferences on training portals like IMPACT [35].

However, further studies point out that the implementation of Telehealth has mostly focussed on training with parents rather than persons with ASD [36],[37]. Heitzman-Powell et al. (2014) [38] have worked on parental training through online learning platform OASIS on knowledge assessment. Bearss et al. [39] have worked on behavioural support through parent training programmes via videoconferencing whereas Suess et al. [40] have worked on parental coaching with didactic training sessions through videoconferences on behaviour principles. Research on parental training on functional communication also include Telehealth coaching at clinical as well as home environments [41].

The above mentioned Telehealth studies on behaviour analysis or development of functional behaviour through various experiments can be broadly categorized into three groups. The first is training of participants either with parents or other interventionists or sometimes on their own with the use of videos and emails. The second set of works is based on the assessment of functional behaviour in a participant through various assignments. The third is the training of parents on ABA for further knowledge based interventions on the participants. The outcomes of these experiments have been to varying degrees. Some are positive, some are not so effective while others could not be validated. These experiments also have different effects on the mental health and behaviour of both the participants and the parents.

Whether the Telehealth based intervention strategies include the effect of emotions, individual preferences and the surrounding environment of a participant in determining one's functional behaviour has not been known. Telehealth practices are advanced strategies which are also difficult to implement in certain situations. They need access to computers and internet which are not accessible in the remotest areas of third world countries like India. There needs to be proper infrastructure for the training of coaches and parents to be able to implement the interventions with fidelity which is again a challenge in countries like India. Finally, the settings in which they are implemented are more experimental than naturalistic although Telehealth researchers claim otherwise.

The current research work takes issue with these technological interventions and strives to create a behavioural model that is easy to understand, easy to work on and creates a naturalistic setting based on minimum interventions with the participants. Considering that the above review works also exclude certain parameters for behavioural measurement, it aims to create a model that covers at least the most important parameters of behavioural experience.

5.3 METHODOLOGY OF THE CASE STUDY

5.3.1 Study Design

The case study was conducted at four centres of NIEPID, India from July, 2016 to September, 2018 in four phases. The first phase was basic observation (classroom behaviour and classroom performance) and screening of the Autistic children. The classroom behaviour observation was about how a child behaves with one's fellow classmates, respective teachers, the helping staffs as well as during activities, tiffin break, while going to the toilet, while coming to the school and at the time of the closing of the school. In case of classroom performance observation, it was seen whether a child accomplishes an activity in the classroom, does daily class work and homework and how one performs in exams.

The second phase consisted of data collection of each child on 1:1 ratio consisting of audio-video data, still images, clinical data (blood pressure, heart rate) and technical data (blood pressure and heart rate recorded through different sensors). Data was also collected from a child's personal record file in the institutions on severity level (borderline, mild, moderate, severe) of one's diagnosis, IQ and associated conditions if any. Care was taken to safely collect the data only for the purpose of research without causing any inconvenience to the children.

The third phase was gathering information of the children about their behavioural patterns from their respective class teachers. The information concerned a child's communication skills, likes and dislikes, tactile aversion, sound sensitivity at the school as well as personal observation of the teachers about each child.

The fourth and final phase was about collecting relevant inputs from the respective caregivers of the children. The information was about a child's communication skills, likes and dislikes, tactile aversion, sound sensitivity at home as well as personal observation of the caregivers about each child.

5.3.2 Participants

The case study enrolled males and females between the age group of 3 to 20 with confirmed diagnosis of Developmental Delay (DD) (within 5 years of age) and Intellectually Disabled (ID) (above 5 years) with associated conditions of ASD.

The study was conducted on the participants [Figure 5.2] with due permission from their caregivers. It should be mentioned that the case study does not include typically developed individuals for comparison with Autistic participants. The study has been done ethically only for the purpose of research following the guidelines of NIEPID.

Figure 5.3 shows the number of male and female students in the four centres. There are total 20 students in NIEPID Secunderabad with 13 males and 7 females. The NIEPID Noida centre consists of 10 male students and no female student. The NIEPID Lajpatnagar

Figure 5.2: Sample Images of all 60 Participants

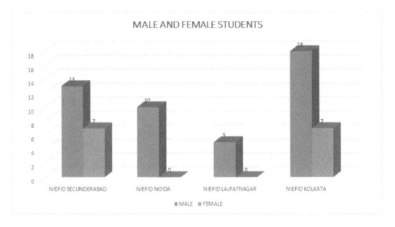

Figure 5.3: Number of Male and Female Students in the Four Centres

centre has 5 male students and no female student. The NIEPID Kolkata centre consists of 25 students with 18 male and 7 female students. Therefore, out of the total 60 students, there are 46 male and 14 female students.

5.3.3 Study Assessment Scale

The study is based on Indian Scale for Assessment of Autism (ISAA) tool which is an evaluation tool for Autistic people based on observation, clinical assessment of behaviour, interaction with the person and collecting information provided by caregivers[42]. ISAA consists of 40 items rated on a scale of 5 from 1 which is always to 5 which is never. The 40 items are further divided into six parts. The first is social relationship and reciprocity. Persons with ASD find it a challenge to communicate with other people. Problems in speech and non-verbal cues such as body language, facial expressions, gestures and postures hinder them to express their thoughts and emotions to others. They prefer not to be bothered and to remain in isolation. The second is emotional responsiveness. Autistic persons experience a wide range of emotions, some of them extremely subtle. They also experience unexpected emotions in a given situation. This unpredictable response may vary from excessive joy to sudden anger and irritation, sometimes without any clear reason. The third part is speech-language and communication. Problems in speech is a common characteristic of Autistic individuals which results in a lack of communication with others. Those who can speak, often use repetitive words, phrases or sentences. Sometimes they also use strange words or sentences with no meaning or context. The fourth is behaviour patterns. People with ASD like to follow routines or patterns and have a dislike for change. They engage in bodily gestures like rocking or making strange sounds or self indulgent behaviour such as flapping hands or playing with certain objects constantly sometimes for no reason but also often to express their current emotional state. The fifth part is sensory aspects. Most Autistic people are either hypersensitive or hyposensitive to light, sound, smell and touch. Some of them explore their surroundings by making use of one or more of their sensory aspects. The sixth and last part is cognitive component. Persons with Autism often lack attention and concentration. However, many of them also have special abilities like singing, drawing, reading, running and so on.

5.3.4 Technical Modules

Different gadgets and sensors have been used to acquire different types of data. For example, blood pressure and heart rate sensors in-built in smart watches have been used to track real time blood pressure and heart rate which have then been stored digitally. A zoom H1 audio recorder has been used to record best quality sounds. A good quality DSLR camera (Canon 700D) has been used for acquiring best quality images (18 mega pixel) and videos (full HD) for the purpose of recognising facial emotions and gestures. Finally, smart phones and tablets have also been used for the monitoring and quick access of data.

5.3.5 Clinical Modules

The clinical aspects of the case study included recording of an Autistic person's blood pressure and heart rate by specially trained staffs. The clinical inputs have then been compared with the

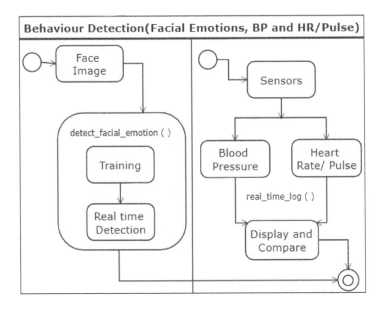

Figure 5.4: Block Diagram of Behaviour Detection through Facial Emotions, BP and Heart Rate

technical inputs to verify whether the sensors have provided accurate data.

Figure 5.4 describes two parallel processes which are detection of facial emotion through camera in real time and collection of blood pressure and heart rate information through sensors. The two parameters are compared with corresponding facial emotions from which behaviour pattern is predicted.

5.4 RESULTS AND DISCUSSION

1. With reference to the first hypothesis it has been found from the case study through NAISI analysis that the likes, dislikes, tactile aversion and sound sensitivity directly affect the behavioural changes of a child either positively or negatively. If a child finds things of one's liking in one's surrounding environment, then one experiences a positive behavioural change and if one finds things to dislike in one's environment, then a hyperactive behavioural change takes place. There is a similar behavioural change with respect to touch and sound.

Figure 5.5 is a screenshot of the NAISI super admin dashboard through which Autistic students are monitored in real time by

Figure 5.5: Super Admin Dashboard of NAISI Portal

administrators, teachers and parents. The teachers can regularly update the children's progress on this portal while the parents can also check their wards' progress.

Table 5.1 below shows a general pattern of likes, dislikes, tactile aversion, sound sensitivity and other observations of the participant children recorded during the case study:

Anything of liking in the surrounding environment of a child, causes a positive behavioural change and vice versa. Children who like ice cream, immediately respond positively if one is given to them. Similarly activities like playing or drawing, singing and so on elicit the right kind of behavioural response from them. This equation also caters to touch and smell. Many Autistic children have a strong dislike for wet and sticky things. If those things are around them or they are given any activity related to those things, then they can start to experience negative behavioural change.

Since every child is different, their tastes are also different. Hence, it has been found during the case study that there are certain things that may be to the liking of a child, but are of a dislike to another. For example, one child may like chocolates while another may dislike them. Giving the second child chocolates will then cause a negative behavioural change. A list of some of the common things and their ratio to both the likes and dislikes of different children is shown in Figure 5.6.

A better knowledge about the likes and dislikes of an Autistic child helps to bring about more awareness around them and improve care of them.

2. With reference to the second hypothesis, Figure 5.7 shows that there are 4 types of behavioural patterns–hyperactive, normal, calm and drowsy. The heart rate of a hyperactive child was found above 90 (the maximum heart rate found in the case study was 118). The heart rate of a child with normal behaviour was found

TABLE 5.1: A General Pattern of Likes, Dislikes, Tactile Aversion, Sound Sensitivity and Other Observations of the Participants

LIKES	• Playing with Items (Toys, Ball, Keys, Shoelaces, Spoons, Teddy Bear) and Spinning them • Playing with Mobile Phones (Playing Video Games) • Playing Badminton and Running • Listening to and Singing Songs, Colouring and Drawing, Dancing, Writing • Watching Television (Cartoons, Cricket) and Videos of Rhymes • Travelling (Roaming Around, Going to Parks, Travelling by Car) • Food (Fast Food, Chips, Cold Drinks, Ice Cream, Chocolate, Vegetables, Vegetable Items and Non-Vegetable Items)
DISLIKES	• Playing with Items (Beads etc) • Playing with Mobile Phones • Listening to Music • Watching Television • Travelling (Especially Going to Crowded Places) and Shopping • Food (Fast Food, Chocolate, Juice, Tonic) • Going to School • Academic Activities and other Short-term Activities • Sitting Near Unhygienic and Wet Places (Washroom) • Weather Sensitive (Especially Rain) • Negative Reply
TACTILE AVERSION	• Human Touch/Touch By Others (Especially Head)/Strangers • Wet, Cold, Sticky, Rough and Dirty Things • Items like Grass, Sand and Dust • Mixing Food With Hands • Dirty Places
SOUND SENSITIVITY	• Loud and Sharp Sounds (Loud Voice, Crackers) • Low Sounds • Loud and Low Sounds • Hearing Problem
OTHER OBSERVATIONS	• Attention Seeking • Throwing Objects • Health Issues and Fits (Under Medication)/Fits (Earlier but Not Now)/Surgery For Fits • Hyperactive (Not Sitting in one place, During Fever)/Lazy • Self-Injurious Behaviour (Banging Head, Biting/Scratching/Pinching Self and Others, Beating Others, Pushing Others) • Speech Delay, Repetition of Words • Crying, Rocking Body, Thumb Sucking, Flapping Hands, Rubbing Ear, Making Teeth Sounds • Needing Repeated Instructions For a Task, Needing Options to Act, Lack of Interest in Learning Methods • Following Instructions and Tasks Completely, Cooperative, High Functioning, Sharp Memory, Adopting Learning Methods Efficiently • Difficulty In Buttoning And Unbuttoning Shirt • Looking at a Person/Thing for a Long Time • Staying at Home • Sexual Activities • Temperature and Smell Sensitivity

Figure 5.6: Likes vs. Dislikes of Common Things

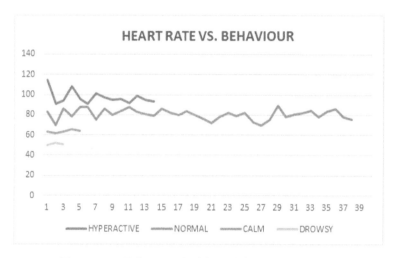

Figure 5.7: Likes vs. Dislikes of Common Things

between 68 and 90. The heart rate of a child with calm behaviour was found between 60 and 67. The heart rate of a child with drowsy behaviour (referring to a few children with additional medication) was found between 57 and 52. Figure 5.7 shows the change of heart rate with change in behaviour.

3. After interviewing the class teachers and the caregivers of each child separately, it has been found that with relation to school, home and activities, the children can be divided into three categories; children who prefer to stay at home and do activities, children who like to come to school but do not like activities and children who like performing activities at school. Figure 5.8 shows that 15 out of 60 children do not like to come to school and prefer their home environment more. They prefer doing activities at home

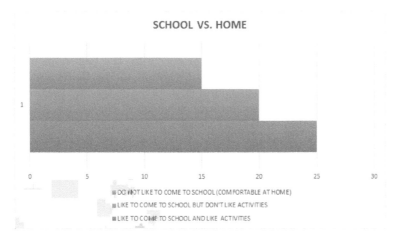

Figure 5.8: Number of Participants who like to come to School and Home

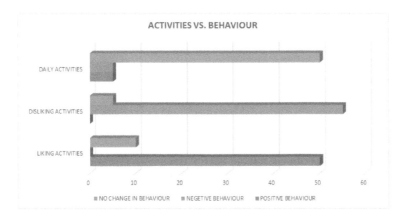

Figure 5.9: Change of Behaviour with respect to Activities

and their behaviour is much more positive there. Twenty out of 60 children like coming to the school but have a dislike for performing activities there. Twenty-five out of 60 children like coming to school and performing activities. They are more approachable, friendly, and cooperative and experience positive behaviour at school.

4. With reference to the last hypothesis it has been found that while performing daily routines or activities, the behaviour of a child is normal whereas as any disruption in the daily routine causes a hyperactive behavioural change. Similarly, any preferred activity of a child causes positive behavioural change whereas if one has to perform an activity to one's dislike, then there is a negative behavioural change. Figure 5.9 below shows the change of behaviour with respect to activities.

5.5 CONCLUSION

The purpose of the work was to predict the behavioural changes of the Autistic children with respect to their blood pressure, heart rate, emotions, likes, dislikes and surrounding environment. The research conducted through the NAISI study at four Autistic learning centres on 60 participant children proved that each parameter shares a certain influence over a child's behaviour. The behavioural model based on the case study provides valuable knowledge and necessary foundation based on which to interact with an Autistic child. Positive communication and encouragement of positive behaviour in an Autistic child can therefore be enforced based on the model. While this was a small study with lesser number of participants and certain necessary parameters of behavioural change, future prospects include case studies with larger number of participants and added parameters to develop the behavioural model further to be able to bring more awareness around Autism and constructing positive and necessary intervention in the lives of the Autistic people.

ACKNOWLEDGEMENT

This work has been funded by SERB (Science and Engineering Research Board). All the data has been collected from NIEPID, India (formerly known as NIMH).

FURTHER READING

[1] American Psychiatric Association, 2013. Diagnostic and statistical manual of mental disorders (DSM-5®). American Psychiatric Pub.

[2] Centers for Disease Control and Prevention. (2018). Autism spectrum disorder (ASD). Retrieved from https ://www.cdc.gov/ncbdd d/autism/data.html.

[3] Begeer, S., Koot, H.M., Rieffe, C., Terwogt, M.M. and Stegge, H., 2008. Emotional competence in children with autism: Diagnostic criteria and empirical evidence. Developmental Review, 28(3), pp.342-369.

[4] Kasari, C., Sigman, M., Mundy, P. and Yirmiya, N., 1990. Affective sharing in the context of joint attention interactions of normal, autistic, and mentally retarded children. Journal of autism and developmental disorders, 20(1), pp.87-100.

[5] Stagg, S.D., Slavny, R., Hand, C., Cardoso, A. and Smith, P., 2014. Does facial expressivity count? How typically developing children respond initially to children with autism. Autism, 18(6), pp.704-711.

[6] McIntosh, D.N., Reichmann-Decker, A., Winkielman, P. and Wilbarger, J.L., 2006. When the social mirror breaks: deficits in automatic, but not voluntary, mimicry of emotional facial expressions in autism. Developmental science, 9(3), pp.295-302.

[7] Grossman, R.B., Edelson, L.R. and Tager-Flusberg, H., 2013. Emotional facial and vocal expressions during story retelling by children and adolescents with high-functioning autism. Journal of Speech, Language, and Hearing Research.

[8] Snow, M.E., Hertzig, M.E. and Shapiro, T., 1987. Expression of emotion in young autistic children. Journal of the American Academy of Child & Adolescent Psychiatry, 26(6), pp.836-838.

[9] Yirmiya, N., Kasari, C., Sigman, M. and Mundy, P., 1989. Facial expressions of affect in autistic, mentally retarded and normal children. Journal of Child Psychology and Psychiatry, 30(5), pp.725-735.

[10] Macdonald, H., Rutter, M., Howlin, P., Rios, P., Conteur, A.L., Evered, C. and Folstein, S., 1989. Recognition and expression of emotional cues by autistic and normal adults. Journal of Child Psychology and Psychiatry, 30(6), pp.865-877.

[11] Volker, M.A., Lopata, C., Smith, D.A. and Thomeer, M.L., 2009. Facial encoding of children with high-functioning autism spectrum disorders. Focus on Autism and Other Developmental Disabilities, 24(4), pp.195-204.

[12] Manfredonia, J., Bangerter, A., Manyakov, N.V., Ness, S., Lewin, D., Skalkin, A., Boice, M., Goodwin, M.S., Dawson, G., Hendren, R. and Leventhal, B., 2019. Automatic recognition of posed facial expression of emotion in individuals with autism spectrum disorder. Journal of autism and developmental disorders, 49(1), pp.279-293.

[13] Langdell, T., 1981. Face perception: An approach to the study of autism (Doctoral dissertation, University of London).

[14] Loveland, K.A., Tunali-Kotoski, B., Pearson, D.A., Brelsford, K.A., Ortegon, J. and Chen, R., 1994. Imitation and expression of facial affect in autism. Development and Psychopathology, 6(3), pp.433-444.

[15] Yoshimura, S., Sato, W., Uono, S. and Toichi, M., 2015. Impaired overt facial mimicry in response to dynamic facial expressions in high-functioning autism spectrum disorders. Journal of autism and developmental disorders, 45(5), pp.1318-1328.

[16] Beadle-Brown, J.D. and Whiten, A., 2004. Elicited imitation in children and adults with autism: Is there a deficit? Journal of Intellectual and Developmental Disability, 29(2), pp.147-163.

[17] Ness, S.L., Bangerter, A., Manyakov, N.V., Lewin, D., Boice, M., Skalkin, A., Jagannatha, S., Chatterjee, M., Dawson, G., Goodwin, M.S. and Hendren, R., 2019. An observational study with the Janssen autism knowledge engine (JAKE®) in individuals with autism spectrum disorder. Frontiers in neuroscience, 13, p.111.

[18] Van Hecke, A.V., Lebow, J., Bal, E., Lamb, D., Harden, E., Kramer, A., Denver, J., Bazhenova, O. and Porges, S.W., 2009. Electroencephalogram and heart rate regulation to familiar and unfamiliar people in children with autism spectrum disorders. Child development, 80(4), pp.1118-1133.

[19] Goodwin, M.S., Intille, S.S., Albinali, F. and Velicer, W.F., 2011. Automated detection of stereotypical motor movements. Journal of autism and developmental disorders, 41(6), pp.770-782.

[20] Goodwin, M.S., Haghighi, M., Tang, Q., Akcakaya, M., Erdogmus, D. and Intille, S., 2014, September. Moving towards a real-time system for automatically recognizing stereotypical motor movements in individuals on the autism spectrum using

wireless accelerometry. In Proceedings of the 2014 ACM International Joint Conference on Pervasive and Ubiquitous Computing (pp. 861-872). ACM.

[21] Großekathöfer, U., Manyakov, N.V., Mihajlović, V., Pandina, G., Skalkin, A., Ness, S., Bangerter, A. and Goodwin, M.S., 2017. Automated detection of stereotypical motor movements in autism spectrum disorder using recurrence quantification analysis. Frontiers in neuroinformatics, 11, p.9.

[22] Aman, M.G., Novotny, S., Samango-Sprouse, C., Lecavalier, L., Leonard, E., Gadow, K.D., King, B.H., Pearson, D.A., Gernsbacher, M.A. and Chez, M., 2004. Outcome measures for clinical drug trials in autism. CNS spectrums, 9(1), pp.36-47.

[23] Aman, M. G., and Singh, N. N. (2017). Aberrant Behavior Checklist Manual, Second Edition. East Aurora, NY: Slosson Educational Publications, Inc.

[24] Constantino, J.N., Davis, S.A., Todd, R.D., Schindler, M.K., Gross, M.M., Brophy, S.L., Metzger, L.M., Shoushtari, C.S., Splinter, R. and Reich, W., 2003. Validation of a brief quantitative measure of autistic traits: comparison of the social responsiveness scale with the autism diagnostic interview-revised. Journal of autism and developmental disorders, 33(4), pp.427-433.

[25] Ferguson, J., Craig, E.A. and Dounavi, K., 2019. Telehealth as a model for providing behaviour analytic interventions to individuals with Autism Spectrum Disorder: a systematic review. Journal of autism and developmental disorders, 49(2), pp.582-616.

[26] Barretto, A., Wacker, D.P., Harding, J., Lee, J. and Berg, W.K., 2006. Using telemedicine to conduct behavioral assessments. Journal of Applied Behavior Analysis, 39(3), pp.333-340.

[27] Barkaia, A., Stokes, T.F. and Mikiashvili, T., 2017. Intercontinental telehealth coaching of therapists to improve verbalizations by children with autism. Journal of applied behavior analysis, 50(3), pp.582-589.

[28] Benson, S.S., Dimian, A.F., Elmquist, M., Simacek, J., McComas, J.J. and Symons, F.J., 2018. Coaching parents to assess and treat self-injurious behaviour via telehealth. Journal of Intellectual Disability Research, 62(12), pp.1114-1123.

[29] Ingersoll, B. and Berger, N.I., 2015. Parent engagement with a telehealth-based parent-mediated intervention program for children with autism spectrum disorders: predictors of program use and parent outcomes. Journal of Medical Internet Research, 17(10), p.e227.

[30] Kuravackel, G.M., Ruble, L.A., Reese, R.J., Ables, A.P., Rodgers, A.D. and Toland, M.D., 2018. Compass for hope: Evaluating the effectiveness of a parent training and support program for children with ASD. Journal of autism and developmental disorders, 48(2), pp.404-416.

[31] Gibson, J.L., Pennington, R.C., Stenhoff, D.M. and Hopper, J.S., 2010. Using desktop videoconferencing to deliver interventions to a preschool student with autism. Topics in Early Childhood Special Education, 29(4), pp.214-225.

[32] Machalicek, W., O'Reilly, M., Chan, J.M., Rispoli, M., Lang, R., Davis, T., Shogren, K., Sorrells, A., Lancioni, G., Sigafoos, J. and Green, V., 2009. Using videoconferencing to support teachers to conduct preference assessments with students with autism and developmental disabilities. Research in Autism spectrum disorders, 3(1), pp.32-41.

[33] Meadan, H., Snodgrass, M.R., Meyer, L.E., Fisher, K.W., Chung, M.Y. and Halle, J.W., 2016. Internet-based parent-implemented intervention for young children with autism: A pilot study. Journal of Early Intervention, 38(1), pp.3-23.

[34] Neely, L., Rispoli, M., Gerow, S. and Hong, E.R., 2016. Preparing interventionists via telepractice in incidental teaching for children with autism. Journal of Behavioral Education, 25(4), pp.393-416.

[35] Ingersoll, B., Wainer, A.L., Berger, N.I., Pickard, K.E. and Bonter, N., 2016. Comparison of a self-directed and therapist-assisted telehealth parent-mediated intervention for children with ASD: A pilot RCT. Journal of Autism and Developmental Disorders, 46(7), pp.2275-2284.

[36] Meadan, H. and Daczewitz, M.E., 2015. Internet-based intervention training for parents of young children with disabilities: A promising service-delivery model. Early child development and care, 185(1), pp.155-169.

[37] Parsons, D., Cordier, R., Vaz, S. and Lee, H.C., 2017. Parent-mediated intervention training delivered remotely for children with autism spectrum disorder living outside of urban areas:

Systematic review. Journal of medical Internet research, 19(8), p.e198.

[38] Heitzman-Powell, L.S., Buzhardt, J., Rusinko, L.C. and Miller, T.M., 2014. Formative evaluation of an ABA outreach training program for parents of children with autism in remote areas. Focus on Autism and Other Developmental Disabilities, 29(1), pp.23-38.

[39] Bearss, K., Burrell, T.L., Challa, S.A., Postorino, V., Gillespie, S.E., Crooks, C. and Scahill, L., 2018. Feasibility of parent training via telehealth for children with autism spectrum disorder and disruptive behavior: A demonstration pilot. Journal of autism and developmental disorders, 48(4), pp.1020-1030.

[40] Suess, A.N., Romani, P.W., Wacker, D.P., Dyson, S.M., Kuhle, J.L., Lee, J.F., Lindgren, S.D., Kopelman, T.G., Pelzel, K.E. and Waldron, D.B., 2014. Evaluating the treatment fidelity of parents who conduct in-home functional communication training with coaching via telehealth. Journal of Behavioral Education, 23(1), pp.34-59.

[41] Lindgren, S., Wacker, D., Suess, A., Schieltz, K., Pelzel, K., Kopelman, T., Lee, J., Romani, P. and Waldron, D., 2016. Telehealth and autism: Treating challenging behavior at lower cost. Pediatrics, 137(Suppl 2), p.S167.

[42] Wong, C.M. and Singhal, S., 2014. INDT-ASD: An autism diagnosis tool for Indian children. Indian Pediatr, 51, pp.355-6.

Index